PROJECT MANAGEMENT CHECKLISTS

*A Complete Guide for
Exterior and Interior Construction*

PROJECT MANAGEMENT CHECKLISTS

*A Complete Guide for
Exterior and Interior Construction*

Fred A. Stitt

VNR VAN NOSTRAND REINHOLD
New York

Copyright © 1992 by Van Nostrand Reinhold

Library of Congress Catalog Card Number 92-4719
ISBN 0-442-01072-9

All rights reserved. No part of this work covered by
the copyright hereon may be reproduced or used in any
form by any means—graphic, electronic, or
mechanical, including photocopying, recording, taping
or information storage and retrieval systems—without
written permission of the publisher.

Printed in the United States of America.

Van Nostrand Reinhold
115 Fifth Avenue
New York, New York 10003

Chapman and Hall
2-6 Boundary Row
London, SE1 8HN, England

Thomas Nelson Australia
102 Dodds Street
South Melbourne 3205
Victoria, Australia

Nelson Canada
1120 Birchmount Road
Scarborough, Ontario M1K 5G4, Canada

16 15 14 13 12 11 10 9 8 7 6 5 4 3 2 1

Library of Congress Cataloging-in-Publication Data

Stitt, Fred A.
    Project management checklists: a complete guide for exterior and
interior construction/Fred Stitt.
      p.   cm.
    ISBN 0-442-01072-9
Includes index.
    1. Construction industry—Management—Handbooks, manuals, etc.
2. Lists.   I. Title.
TH483.S75 1992
690'.068 —dc20   92-4719
                                                         CIP

CONTENTS

	INSTRUCTIONS	vii
PHASE 1:	PREDESIGN	
	MARKETING AND PRESENTATION MANAGEMENT	1
	FEASIBILITY AND FINANCIAL ANALYSIS	4
	PRECONTRACTUAL ADMINISTRATION	8
	PROJECT PLANNING	14
	PROGRAMMING AND PREDESIGN	17
PHASE 2:	SITE ANALYSIS	
	PREDESIGN AND SCHEMATIC SITE REVIEW	28
	ENVIRONMENTAL IMPACT REPORT	30
	PERMITS AND APPROVALS	32
PHASE 3:	SCHEMATIC DESIGN	
	CONSTRUCTION COST ESTIMATING	37
	SCHEMATIC DESIGN AND DOCUMENTATION	39
	BUILDING CODE AND FIRE CODE SEARCH	49
PHASE 4:	DESIGN DEVELOPMENT	59
PHASE 5:	CONSTRUCTION DOCUMENTS	
	WORKING DRAWINGS	72
	CONSULTANT/ENGINEERING DRAWINGS CROSS-COORDINATION CHECKLIST	86
	SPECIFICATION WRITING AND COORDINATION	94
PHASE 6:	PREBIDDING, BIDDING AND NEGOTIATIONS	100
PHASE 7:	CONSTRUCTION CONTRACT ADMINISTRATION	108
	SHOP DRAWING CHECKING AND COORDINATION	123
PHASE 8:	POSTCONSTRUCTION ADMINISTRATION	126
PHASE 9:	LONG RANGE MARKETING PLANNING	130

INSTRUCTIONS

This book is the most comprehensive architectural project management checklist ever published.

It includes checklists of all the primary tasks required in managing A/E projects of every type and size. The checklists are subdivided according to all the main contract phases of A/E work from Predesign through Post Construction Administration and beyond.

And the lists have multiple uses and values for management and for preventing memory lapses and misunderstandings that lead to the claims, counterclaims, and lawsuits that plague our industry.

Here are some of the applications of this book:

1) Each checklist is a personal work planner -- a project management "To Do" list.

2) Items that are checked as things to do will provide reminders for later back checking of what's been completed or not completed along the way.

3) The lists are step-by-step instruction guides for younger personnel on how experienced project managers organize their work.

4) When work is delegated or shared, the checklists serve as an ongoing coordination reminder to everyone to do what they agreed to when they agreed to.

5) As work is completed, the filled in checklists provide a detailed history of project decisions and actions. This is what's needed to document changes in the Scope of Work and for resolving disputes that arise later when people forget when and why decisions and changes were made along the way.

Remember to treat this original checklist as a permanent office resource; make photocopies of individual chapters for your own use as you need them. And drop me a line if you have any suggestions for improving any of the checklists.

Fred Stitt, Editor/Publisher, GUIDELINES.
Box 456, Orinda, CA 94563. FAX (510) 254-9397.

PROJECT MANAGEMENT CHECKLISTS

A Complete Guide for
Exterior and Interior Construction

Phase 1: Predesign 1
Marketing & Presentation Management

PROJECT MANAGEMENT CHECKLIST

Project Name/No: Notes by:

Dates Checked:

This checklist section, PROJECT PRESENTATION PLANNING is an operational To-Do list to use on a project by project basis.

A related checklist -- LONG RANGE MARKETING PLANNING -- provided at the end of the book, is an administrative review for managing all general marketing activities.

PROJECT PRESENTATION PLANNING

Checkmark each item to be done and cross out the check when completed. Mark with a -- if an item is not to be done. If an item is in doubt, mark with question mark and add a note of what to do to resolve the question. By: Dates:

___ Identify who is in charge and/or what is the source of each aspect of the client presentation.

   ___ Administration.   _____
   ___ Presentation planning and storyboard.   _____
   ___ Written text or narration.   _____
   ___ Photos/slides.   _____
   ___ Sketches/renderings.   _____
   ___ Models.   _____
   ___ Video recording/editing.   _____
   ___ Tape recording.   _____
   ___ Production equipment.   _____
   ___ Presentation equipment.   _____
   ___ Client presentation.   _____
   ___

___ Verify that all stated client requests and requirements have been addressed prior to presentation.

___ If it is clear that requests of the client won't be fully satisfied at the formal presentation, prepare material to show the reasons why.

___ Plan a scenario of what's to be shown in the presentation.

   ___ Review of the problems and requirements of the project.
   ___ Description of the method and process of solving the problems.
   ___ Review of alternative solutions.
   ___ The criteria for judging alternative solutions.
   ___ The final proposed solution(s).
   ___

Add notes as needed after each checklisted task, such as:
Initials of who is to do it, when it's to start, when to review, who to coordinate with, and when it's to be finished.

Phase 1: Predesign 2
Marketing & Presentation Management

PROJECT PRESENTATION PLANNING continued

Checkmark each item to be done and cross out the check when completed.
If an item is in doubt, mark with question mark and add a note of what to do to resolve the question. By: Dates:

___ Decide on the presentation media best suited to the client and the project.

    ___ Drawings and renderings:
        ___ Conceptual sketches/process drawings.
        ___ Spatial allocation charts, link and node, and bubble diagrams.
        ___ Finish renderings.
        ___ Combination photo-drawings.
        ___

    ___ Text:
        ___ Written outline of the design process: Requirements/Problem Solving Process/Criteria/Solution(s).
        ___ Computer printout analysis.
        ___ Feasibility report.
        ___

    ___ Photos and/or Slides:
        ___ Multiple or single-screen slides.
        ___ Front or back projection.
        ___ Photos of computer screen images.
        ___ Photos of sketches and process drawings.
        ___ Photos of planning charts and diagrams.
        ___ Site survey and site surroundings.
        ___ Combination of site and drawing or model.
        ___ Back-projected slides.
        ___ Projected simulation of full-size spaces.
        ___ Slides with taped narrative.
        ___

    ___ Models:
        ___ Study models, small scale.
        ___ Study models, large scale.
        ___ Partial full-size mockups.
        ___

    ___ Overhead projector:
        ___ Photos.
        ___ Drawings and charts.
        ___ On the spot drawing.
        ___

    ___ Video:
        ___ (Video can include virtually all the items just listed.)

Add notes as needed after each checklisted task, such as:
Initials of who is to do it, when it's to start, when to review, who to coordinate with, and when it's to be finished.

Phase 1: Predesign 3
Marketing & Presentation Management

PROJECT PRESENTATION PLANNING continued

Checkmark each item to be done and cross out the check when completed.
If an item is in doubt, mark with question mark and add a note of what to do to resolve the question. By: Dates:

___ Review presentation participants, their speaking roles, and limits of their roles.

___ Identify all design study materials that can be reused as displays or photo/video visuals in the presentation.

   ___ Conceptual sketches.
   ___ Charts/diagrams.
   ___ Study models.
   ___ Computer display screen images.
   ___

___ Identify design and presentation materials that can be adapted and reused in later client graphics.

   ___ Project publicity and press releases.
   ___ Building sales/rental brochure.
   ___ Signage/direction maps.
   ___ Building maintenance and facilities management key plans.
   ___

___ Identify all project documents and graphics that can be reused in the long term office marketing program. Arrange transfer of reusable design and production graphics to the marketing director.

   ___ Publicity/press releases.
   ___ News released to professional journals.
   ___ Office brochures.
   ___

___ Identify presentation materials that can be reused in future presentations. Notify the marketing director.

___ Identify all design study and presentation graphics that can be reused in production drawings. Arrange for transfer of reusable design graphics to the production manager.

   ___ Site survey.
   ___ Site study photographs.
   ___ Finish presentation plans and elevations as base and overlay sheets so base sheets can be reused in working drawings.
   ___ Coordinate drawing sizes, orientation and scale with the working drawing production department.
   ___ Free hand design sketch details and sections.
   ___

___ Schedule call backs to the client to review additional design service needs.

___ Request referrals from client of others who might be in need of the office's design services.

Add notes as needed after each checklisted task, such as:
Initials of who is to do it, when it's to start, when to review, who to coordinate with, and when it's to be finished.

Phase 1: Predesign 4
Feasibility and Financial Analysis

PROJECT MANAGEMENT CHECKLIST

Project Name/No: Notes by:

Dates Checked:

Conduct a feasibility and financial analysis of every proposed project for evidence that it is financially and politically realistic. If this is not desired by the client as a part of designated services, then it should still be done to some degree solely for the benefit of the design firm prior to the execution of the design contract. The worst losses for many design firms come from non-payment for work executed on unrealistic projects.

The first portion of this checklist, PROJECT FEASIBILITY, includes elements to consider in a general feasibility study. The second portion, FINANCIAL AND CASH FLOW ANALYSIS, deals with the financial aspects that will guide lenders and investors in supporting the project.

PROJECT FEASIBILITY

Checkmark each item to be done and cross out the check when completed. Mark with a -- if an item is not to be done. If an item is in doubt, mark with question mark and add a note of what to do to resolve the question. By: Dates:

___ Market review, subjective and analytical.

    ___ Location/site ratings (Positive, Median, Negative).

        ___ Social/economic.
        ___ Growth/area property improvements.
        ___ Climate/microclimate.
        ___ Solar orientation.
        ___ Views.
        ___ Transportation.
        ___ Parking.
        ___ Support services.
        ___ Security.
        ___ Proximity of similar functions.
        ___

___ Identify the client's and prospective lenders' weightings of the importance of the previously listed aspects of the site location. Do they coincide or differ?

___ Legal and regulatory considerations. Create separate lists as necessary.

        ___ Variances required.
        ___ Special permits.
        ___ Special interest or neighborhood group opposition.
        ___ Deed restrictions.
        ___ Possible future rent or use controls.
        ___ Possible retroactive regulations or controls.
        ___

Add notes as needed after each checklisted task, such as:
Initials of who is to do it, when it's to start, when to review, who to coordinate with, and when it's to be finished.

Phase 1: Predesign 5
Feasibility and Financial Analysis

PROJECT FEASIBILITY   continued

Checkmark each item to be done and cross out the check when completed. Mark with a -- if an item is not to be done. If an item is in doubt, mark with question mark and add a note of what to do to resolve the question.   By: Dates:

___ Time schedule estimates.

    ___ Permit process.
    ___ Financing.
    ___ Design and documentation.
    ___ Construction.
    ___ Rental or sales to the break-even point.
    ___

    ___ Special financial possibilities. Make separate lists as necessary.
    ___ Historical preservation grant or facade easement.
    ___ Value enhancement by rezoning.
    ___ Property tax moratorium or reduction.
    ___ Grants/subsidies.
    ___ FHA or other agency-insured financing.
    ___ Guaranteed loans.
    ___ Tax-exempt revenue bond financing.
    ___ Leaseback/lease financing.
    ___ Tax credits and deductions.
    ___ Nonprofit institutional participation.
    ___

    ___ Real estate market analysis. Make separate lists as necessary.

    ___ Absorption Rate. (The units successfully marketed in the previous year compared to the supply.)

    ___ Capture Rate. (The ratio of units to be supplied by the project to the total units absorbed per year.)

    ___ Market Rent. (Current and projected unit income for comparable projects in the market area.)

    ___ Other similar projects--current occupancy rates.

    ___ Other similar projects--current new planning.

    ___ Cost and income projections (see the next page).

    ___

___ Review and evaluate the client's creditworthiness and financing resources.

___ Review and evaluate the client's need and commitment to complete the proposed project.

Add notes as needed after each checklisted task, such as:
Initials of who is to do it, when it's to start, when to review, who to coordinate with, and when it's to be finished.

Phase 1: Predesign 6
Feasibility and Financial Analysis

FINANCIAL AND CASH FLOW ANALYSIS

Checkmark each item to be done and cross out the check when completed. Mark with a -- if an item is not to be done. If an item is in doubt, mark with question mark and add a note of what to do to resolve the question. By: Dates:

___ Estimate the project's total construction cost and financing costs.

    ___ Property cost.   _____
    ___ Survey and soil tests.
    ___ Site preparation.
    ___ Predesign and programming fees.
    ___ Architectural fees.
    ___ Engineering fees.
    ___ Other consultant's fees.
    ___ Permits, testing, and inspection fees.
    ___ Tenant space planning.
    ___ Furniture.
    ___ Equipment.
    ___ Landscaping.
    ___ Legal fees.
    ___ Property taxes during construction.
    ___ Insurance during construction.
    ___ Mortgage loan fees.
    ___ Interim loan fee.
    ___ Interest.
    ___ Title and closing costs.
    ___ Postconstruction design services.
    ___ Leasing agent fees.
    ___ Contingency allowance.
___

___ Estimate the building's debt service and operating costs per square foot.

    ___ Debt service.
    ___ Utilities:
        ___ Lighting and power.
        ___ Gas and/or fuel.
        ___ Tenant communications services.
    ___

    ___ Facilities management.
    ___ Leasing management.
    ___ Cleaning.
    ___ Maintenance and repairs.
    ___ Painting and decorating.
    ___ Landscape and exterior maintenance.
    ___ Property taxes.
    ___ Other taxes.

Add notes as needed after each checklisted task, such as:
Initials of who is to do it, when it's to start, when to review, who to coordinate with, and when it's to be finished.

Phase 1: Predesign 7
Feasibility and Financial Analysis

FINANCIAL AND CASH FLOW ANALYSIS   continued

Checkmark each item to be done and cross out the check when completed. Mark with a -- if an item is not to be done.
If an item is in doubt, mark with question mark and add a note of what to do to resolve the question.   By: Dates:

___ Insurance.
___ Accounting fees.
___ Security.
___ Inflation index.
___ Contingency allowance.

___ Estimate the building's likely gross rental, lease, or sale income:

   ___ Space or function types.
   ___ Square footage.
   ___ Income per square foot.
   ___ Vacancy rate.

___ Compare estimated occupancy rate, income, and costs annually to establish break-even point and future income production--a Cash Flow Statement.

   ___ Occupancy percentage annually or semi-annually.
   ___ Annual or semiannual income.
   ___ Annual or semiannual expenses.
   ___ Percentage of operating expenses met by income.
   ___ Amount of income remaining for debt service.
   ___ Remaining annual cash flow.

___ Estimate tax considerations for owners, tenants, and investors.

   ___ Investment tax credit.
   ___ Building depreciation.

      ___ Straight line.
      ___ Declining balance.
      ___ Sum of the years.

   ___ Depreciation on furnishings and equipment.
   ___ Deferred taxes on unrealized depreciation.
   ___ Interest deductions.
   ___ Capital gain tax rate on profit from resale.

Add notes as needed after each checklisted task, such as:
Initials of who is to do it, when it's to start, when to review, who to coordinate with, and when it's to be finished.

Phase 1: Predesign 8
Precontractual Administration

PROJECT MANAGEMENT CHECKLIST

Project Name/No: Notes by:

Dates Checked:

These tasks are required before final execution of the design services contract with the client. Marketing and Presentation Management precedes this stage and overlaps it to a degree.

CLIENT QUALIFICATION, PROGRAM AND BUDGET REVIEW

Checkmark each item to be done and cross out the check when completed. Mark with a -- if an item is not to be done. If an item is in doubt, mark with question mark and add a note of what to do to resolve the question.   By: Dates:

___ Verify the financial qualifications of the client.

___ Review the previous project track record of client.

___ Identify the actual movers and decision makers in the client organization and evaluate their capabilities and likely compatibility with the design team.

___ Examine the client's budget estimates and determine if the budget includes reasonable allowances for:

   ___ Land acquisition.
   ___ Predesign and programming services.
   ___ Sitework.
   ___ Special foundation work.
   ___ Surveys, tests, and inspections.
   ___ Demolition.
   ___ New construction.
   ___ Client's in-house management and staff costs.
   ___ Remodeling.
   ___ Equipment.
   ___ Furnishings.
   ___ Landscaping.
   ___ Architects' and consultants' fees.
   ___ Legal fees.
   ___ Finance costs.
   ___ Insurance.
   ___ Taxes.
   ___ Escalation.
   ___ Contingency.
   ___

Add notes as needed after each checklisted task, such as:
Initials of who is to do it, when it's to start, when to review, who to coordinate with, and when it's to be finished.

Phase 1: Predesign 9
Precontractual Administration

PROGRAMMING REVIEW

Checkmark each item to be done and cross out the check when completed. Mark with a -- if an item is not to be done.
If an item is in doubt, mark with question mark and add a note of what to do to resolve the question. By: Dates:

___ Evaluate the adequacy of the client's program.

___ Review the client's project time schedule.

___ Evaluate whether the client's program, budget, and schedule are realistic and coordinated with one another.

PROJECT FINANCIAL AND MARKET FEASIBILITY REVIEW

Checkmark each item to be done and cross out the check when completed. Mark with a -- if an item is not to be done.
If an item is in doubt, mark with question mark and add a note of what to do to resolve the question. By: Dates:

___ Verify the financial viability of the project: costs vs. projected income.

___ Review the local marketability of the project.

PERMITS AND APPROVAL REVIEW

Checkmark each item to be done and cross out the check when completed. Mark with a -- if an item is not to be done.
If an item is in doubt, mark with question mark and add a note of what to do to resolve the question. By: Dates:

___ Evaluate the likely sources and strength of opposition to the project.

   ___ Sources:

   ___ Strength:
      ___ Numerical.
      ___ Media support.
      ___ Financial/legal.
      ___ Political.
      ___

   ___ Likely opposition strategies.

___ Review fallback positions, trade-offs, and the effect on financial feasibility of possible compromises with project opponents.

Add notes as needed after each checklisted task, such as:
Initials of who is to do it, when it's to start, when to review, who to coordinate with, and when it's to be finished.

Phase 1: Predesign 10
Precontractual Administration

SCOPE OF WORK/TYPE OF CONTRACT

Checkmark each item to be done and cross out the check when completed. Mark with a -- if an item is not to be done. If an item is in doubt, mark with question mark and add a note of what to do to resolve the question. By: Dates:

___ Determine what will constitute Basic Services as opposed to Additional Services in Owner-Architect Agreement.

___ Identify all major special or Additional Services that may be required (AIA Doc. B727).

    ___ Construction management.
    ___ Long term master planning.
    ___ Special energy conservation or solar design and engineering.
    ___ Permits and approvals campaign.
    ___ Project financing campaign and/or participation.
    ___ Interior design and furnishings.
    ___ Sales, leasing, and tenant management.
    ___ Facilities maintenance management.
    ___

___ Determine the likely or decided method of awarding the construction contract:

    ___ Competitive Bidding -- Open.
    ___ Competitive Bidding -- Selected Contractors.
    ___ Negotiated Contract.
    ___ Single Prime Contract.
    ___ Multiple Separate Contracts.
    ___ Stipulated Lump Sum (AIA Doc. A101, A107).
    ___ Cost Plus Fee (AIA Doc. A111, A117).
    ___

___ Contractor Fee types:

    ___ Fixed Fee.
    ___ Fixed Fee with Guaranteed Maximum.
    ___ Percentage of Construction.
    ___

___ Related options:

    ___ Phased Construction.
    ___ Fast Track.
    ___ Construction Management.
    ___ Design-Build.
    ___ Contractor prepared construction documents.
    ___

Add notes as needed after each checklisted task, such as:
Initials of who is to do it, when it's to start, when to review, who to coordinate with, and when it's to be finished.

Phase 1: Predesign 11
Precontractual Administration

CONSULTANT PREPARATION AND COORDINATION

Checkmark each item to be done and cross out the check when completed. Mark with a -- if an item is not to be done.
If an item is in doubt, mark with question mark and add a note of what to do to resolve the question. By: Dates:

___ Review and reach agreement on conditions of contracts with consultants.

      ___ Fees and method of compensation.
      ___ Work schedule.
      ___ Licensing for the project locale.
      ___ Liability and other required insurance.
      ___

___ Organize the team (structural, mechanical, electrical, civil, and any special consultants) and negotiate tentative compensation. Verify all consultants' abilities to meet the time schedule, liability insurance, and licensing requirements for the project.

OFFICE CAPABILITY REVIEW

Checkmark each item to be done and cross out the check when completed. Mark with a -- if an item is not to be done.
If an item is in doubt, mark with question mark and add a note of what to do to resolve the question. By: Dates:

___ Confirm the office facilities, managerial, and personnel capabilities for the project.

      ___ Conflicts with other scheduled projects.
      ___ Required new hiring.
      ___ Required expansion.
      ___

___ Verify special regional liability and licensing requirements. (AIA Doc. HBC 4).

___ Identify specialized technical joint venture or consultant assistance that might be needed.

___ Identify special regional joint venture or consultant assistance that might be needed.

___ Identify any special branch office facilities that might be required.

      ___ Location.
      ___ Cost of facility.
      ___ Cost of temporary or regional personnel.
      ___ Travel and communication costs between home office and branch office(s).
      ___

Add notes as needed after each checklisted task, such as:
Initials of who is to do it, when it's to start, when to review, who to coordinate with, and when it's to be finished.

Phase 1: Predesign 12
Precontractual Administration

AGREEMENT/CONTRACT TERMS

Checkmark each item to be done and cross out the check when completed. Mark with a -- if an item is not to be done. If an item is in doubt, mark with question mark and add a note of what to do to resolve the question.   By: Dates:

___ Do worksheets to record estimated design service time and costs. (AIA Doc. F721).

    ___ Estimated work time required for Basic Services.
    ___ Estimated work time required for Additional Services.
    ___ Estimated total cost of Basic Services.
    ___ Estimated total cost of Additional Services.
    ___ Estimated consultant fees and coordination costs.
    ___

    ___ Estimated reimbursable costs:

        ___ Travel.
        ___ Communications.
        ___ Reprographics.
        ___ Computer services.
        ___ Special consultants.
        ___ Presentation materials.
        ___

___ Method of compensation for Basic Services.

    ___ Percentage of construction cost.
    ___ Hourly cost plus multiplier.
    ___ Cost plus profit.
    ___ Cost plus with guaranteed maximum.
    ___ Fixed fee.
    ___

___ Method of compensation for Additional Services.

    ___ As per Basic.
    ___ Negotiated fixed fees.
    ___ Cost plus.
    ___

___ Estimated total work times in estimated work hours.
___ Work hour costs.
___

___ Review all conditions of the Owner-Architect Agreement. Consult the office attorney regarding any unusual provisions.

Add notes as needed after each checklisted task, such as:
Initials of who is to do it, when it's to start, when to review, who to coordinate with, and when it's to be finished.

Phase 1: Predesign 13
Precontractual Administration

AGREEMENT/CONTRACT TERMS continued

Checkmark each item to be done and cross out the check when completed. Mark with a -- if an item is not to be done. If an item is in doubt, mark with question mark and add a note of what to do to resolve the question. By: Dates:

___ Complete the Owner-Architect Agreement and submit it to the client for approval and acceptance.

___ Distribute memos on any special contract terms to concerned parties for their approval and/or comment.

    ___ Principals/Associates.
    ___ Department heads.
    ___ Consultants.
    ___ Affiliated firms.
    ___

___ Review and negotiate any special contract terms proposed by the client.

___ Conduct a final review of accepted contract terms with the attorney.

___ Confirm that whoever is signing for the client is the client's authorized legal agent.

___ Complete the signing and initialing of the contract and schedule or initiate the first formal phase of work.

___ Distribute copies of the final agreement, or pertinent sections, to all concerned parties.

Add notes as needed after each checklisted task, such as:
Initials of who is to do it, when it's to start, when to review, who to coordinate with, and when it's to be finished.

Phase 1: Predesign 14
Project Planning

PROJECT MANAGEMENT CHECKLIST

Project Name/No: Notes by:

Dates Checked:

If using this checklist in a planning meeting, mark each task to do, write initials of who is to do it, when the task should be reviewed, who to coordinate with, and when it's to be completed.

PROJECT PLANNING AND ADMINISTRATION

Checkmark each item to be done and cross out the check when completed. Mark with a -- if an item is not to be done. If an item is in doubt, mark with question mark and add a note of what to do to resolve the question.

NOTE: The steps listed here are taken after the contract with the client has been executed. This is the administrative preparation that's required as the office proceeds into predesign work, programming, and site analysis.

___ Identify the next immediate steps and phases as provided in the Owner-Architect Agreement Scope of Work:

    ___ Feasibility/market studies.
    ___ Financial feasibility.
    ___ Assistance in financing applications.
    ___ Agency/commission permit applications.
    ___ Environmental impact report.
    ___ Programming.

___ Identify the fee available for programming, schematics, and design development. Compare the available fee with the office's hourly costs, and establish the number of work hours that can be allotted to each phase and each major subdivision of the phases.

___ Create job calendar of estimated phase starts and completions, and phase payments based on monthly installments of estimated final fee.

    ___ Predesign and Programming.
        ___ Site Analysis.
        ___ Schematic Design.
        ___ Design Development.
        ___ Construction Documents.
        ___ Bidding/Negotiation.
        ___ Contract Administration.
        ___ Post-Construction

___ Distribute copies of the job calendar to all job participants.

Add notes as needed after each checklisted task, such as:
Initials of who is to do it, when it's to start, when to review, who to coordinate with, and when it's to be finished.

Phase 1: Predesign 15
Project Planning

PROJECT PLANNING AND ADMINISTRATION continued

Checkmark each item to be done and cross out the check when completed.
If an item is in doubt, mark with question mark and add a note of what to do to resolve the question.

___ Create a schedule for budget and schedule progress reviews (AIA Doc. F721, F723, F800 Series).

___ Create a list of contents based on this Project Management Checklist for your job diary/project record book.

___ Establish a project accounting record format.

___ Establish dates for preparation of monthly expense records (AIA Doc. F5002).

___ Establish dates for submittal of statements to client.

___ Establish dates for preparation of monthly reimbursable expense statements.

___ Assign or confirm the project identification number.

___ Create categories for the project filing system.

___ Prepare file folders and/or binders for project records.

___ Create project data forms (AIA Doc. G809).

___ Create the project directory (AIA Doc. G807).

___ Acquire from the client the name(s), address(es), phone number(s) and authority of the client representative(s). Record all names in the project directory.

PERSONNEL ALLOCATION

Checkmark each item to be done and cross out the check when completed. Mark with a -- if an item is not to be done.
If an item is in doubt, mark with question mark and add a note of what to do to resolve the question. By: Dates:

___ Identify and assign the appropriate project staff well in advance of project phase start dates.

___ Confirm staff member assignments with the personnel manager and add the new project assignments to the office's project record chart.

Add notes as needed after each checklisted task, such as:
Initials of who is to do it, when it's to start, when to review, who to coordinate with, and when it's to be finished.

Phase 1: Predesign 16
Project Planning

CONSULTANT COORDINATION

Checkmark each item to be done and cross out the check when completed. Mark with a -- if an item is not to be done. If an item is in doubt, mark with question mark and add a note of what to do to resolve the question.   By: Dates:

___ Obtain verification of professional liability coverage from all consultants.

___ Complete contractual agreements with consultants. Notify the client of all consultant choices and confirm the client's approval.

___ Send copies of relevant portions of the Owner-Architect Agreement (particularly those pertaining to the Scope of Work) to project consultants.

___ Identify all consultant and client contract sections that might be useful to project participants and distribute them accordingly.

> ___ Competitive Bidding -- Open.
> ___ Competitive Bidding -- Selected Contractors.
> ___ Negotiated Contract.
> ___ Single Prime Contract.
> ___ Multiple Separate Contracts.
> ___ Stipulated Lump Sum.
> ___ Cost Plus Fee.

> Fee Types:
>
> > ___ Fixed Fee.
> > ___ Fixed Fee with Guaranteed Maximum.
> > ___ Percentage of Construction.
> > ___ Hourly.
> > ___ Square Footage.

> Related options:
>
> > ___ Phased Construction.
> > ___ Fast Track.
> > ___ Construction Management.
> > ___ Design-Build.
> > ___ Contractor Prepared Construction Documents.
> > ___ Data Base Facilities Management.

Add notes as needed after each checklisted task, such as:
Initials of who is to do it, when it's to start, when to review, who to coordinate with, and when it's to be finished.

Phase 1: Predesign 17
Programming and Predesign

PROJECT MANAGEMENT CHECKLIST

Project Name/No: Notes by:

Dates Checked:

SCOPE OF PROGRAMMING SERVICE

Checkmark each item to be done and cross out the check when completed. Mark with a -- if an item is not to be done. If an item is in doubt, mark with question mark and add a note of what to do to resolve the question. By: Dates:

___ Determine whether the creation of a building design program is part of the designated services required by the client.

___ Determine whether the client will provide a complete program.

___ Determine whether the client will provide a partial program, with completion to be done by others, such as tenants.

___ Determine whether facilities consultant(s) will provide a complete or partial program.

DATA GATHERING FROM CLIENT

Checkmark each item to be done and cross out the check when completed. Mark with a -- if an item is not to be done. If an item is in doubt, mark with question mark and add a note of what to do to resolve the question. By: Dates:

___ Obtain the client's list of building functions and spaces.

___ Obtain the client's list of equipment and furnishings.

___ Obtain the client's building construction and operating cost estimates. (See the section on FEASIBILITY AND FINANCIAL ANALYSIS.)

___ Proceed with client and user surveys. (See the questionnaires at the end of this division.)

Add notes as needed after each checklisted task, such as:
Initials of who is to do it, when it's to start, when to review, who to coordinate with, and when it's to be finished.

Phase 1: Predesign 18
Programming and Predesign

PROGRAMMING -- BUILDING CONFIGURATION, CONSTRUCTION, AND MATERIALS

Checkmark each item to be done and cross out the check when completed. Mark with a -- if an item is not to be done.
If an item is in doubt, mark with question mark and add a note of what to do to resolve the question. By: Dates:

___ Identify overall occupancy and specific departmental and room occupancies.

___ Verify the client's occupancy and spatial estimates. Identify possible client errors and omissions for later client review.

___ Obtain lists:

    ___ Departments and relationships to other departments.
    ___ Building rooms and relationship to other rooms.
    ___ Occupancies--type and number.
    ___ Equipment and equipment functions.
    ___ Special furnishings requiring custom design/fabrication.
    ___ Owner-supplied equipment and furnishings.

___ List required or optional provisions for phased construction and future additions.

___ Identify property building line limitations to estimate ground level building area.

___ Verify site zoning or other restrictions on building height.

___ Identify orientation considerations:

    ___ Climatic.
    ___ Energy.
    ___ Views.
    ___ Traffic/parking.
    ___ Public transportation.
    ___ Regulatory or deed restrictions.

___ Identify options of numbers of building stories and total height based on estimated floor plan areas and overall occupancy.

___ Estimate size(s) of core area(s) required for:

    ___ Mechanical services.
    ___ Electrical chases.
    ___ Vertical transportation.
    ___ Stairs/smoke towers.

Add notes as needed after each checklisted task, such as:
Initials of who is to do it, when it's to start, when to review, who to coordinate with, and when it's to be finished.

Phase 1: Predesign 19
Programming and Predesign

PROGRAMMING -- BUILDING CONFIGURATION, CONSTRUCTION, AND MATERIALS
continued

Checkmark each item to be done and cross out the check when completed. Mark with a -- if an item is not to be done.
If an item is in doubt, mark with question mark and add a note of what to do to resolve the question. By: Dates:

___ Estimate structural spans required to suit room spatial needs.

___ Identify options of structural systems.

___ Identify options of building configuration based on functions, occupancies, site limitations, orientation, height, spans, and structural system.

___ Identify options of construction systems suited to the likely building configuration and structural system.

___ Identify building cladding and fenestration suited to construction, structural, functional, and cost considerations.

___ Identify interior partitioning, flooring, and ceiling systems suited to construction, structural, functional, and cost considerations.

___ Estimate construction and site development construction costs.

___ Establish who identifies room relationships and their importance.

    ___ An averaging of general user estimates.
    ___ Client management.
    ___ Facilities management.
    ___ Facilities planning consultant.
    ___ Department heads.
    ___ User groups or individuals.
    ___ Tenants.
    ___

PROGRAMMING -- OCCUPANCY NEEDS AND SPATIAL ALLOCATION

Checkmark each item to be done and cross out the check when completed. Mark with a -- if an item is not to be done.
If an item is in doubt, mark with question mark and add a note of what to do to resolve the question. By: Dates:

___ Establish criteria for importance of room functions and relationships, and create a User Questionnaire. (See sample at the end of this division.)

    ___ Volume of traffic.
    ___ Frequency of interaction.
    ___ Relative value or cost of the interactions or personnel.
    ___

___ Create a Departmental Spatial Interaction Matrix (list of departments that shows their relationship to other departments).

Add notes as needed after each checklisted task, such as:
Initials of who is to do it, when it's to start, when to review, who to coordinate with, and when it's to be finished.

Phase 1: Predesign 20
Programming and Predesign

PROGRAMMING -- OCCUPANCY NEEDS AND SPATIAL ALLOCATION continued

Checkmark each item to be done and cross out the check when completed. Mark with a -- if an item is not to be done.
If an item is in doubt, mark with question mark and add a note of what to do to resolve the question. By: Dates:

___ Create room by room spatial interaction diagrams showing all room relationships.

___ Identify numerical ratings of the importance of relationships of each room to other rooms.

___ Make link and node diagrams to show departmental and room relationships identified in the interaction matrices.

___ Make bubble diagrams indicating spaces with relationships and their importance rankings. Manipulate bubble diagrams until link crossovers (plan conflicts) are eliminated.

___ Create diagrammatic/schematic building plans.

___ Note relative spatial areas for all departments, rooms, mechanical, vertical transportation, service, exit stairs and corridors, and horizontal circulation.

___ Review program and predesign decisions with client.

THE PRELIMINARY DESIGN PROGRAM CHECKLIST

Checkmark each item to be done and cross out the check when completed. Mark with a -- if an item is not to be done.
If an item is in doubt, mark with question mark and add a note of what to do to resolve the question. By: Dates:

The following checklist identifies and consolidates the basic predesign information needed to get the project under way.

The first phase of predesign is broad; it establishes design attributes and limitations in words and numbers.

Later predesign decision-making sessions will cycle downward in smaller and smaller units--from the building as a whole, to subdivisions, to individual rooms, to individual elements and features of each room.

The basic attribute identification process in checklist format:

___ Client.

___ Client representatives (Names, titles, addresses, phone).

___ Chain of responsibility or decision making in client's organization.

___ Source of financing.

___ Client's generally stated needs and desires.

___ Overriding goal/purpose of building project.

___ Primary building functions.

Add notes as needed after each checklisted task, such as:
Initials of who is to do it, when it's to start, when to review, who to coordinate with, and when it's to be finished.

Phase 1: Predesign 21
Programming and Predesign

THE PRELIMINARY DESIGN PROGRAM CHECKLIST continued

Checkmark each item to be done and cross out the check when completed. Mark with a -- if an item is not to be done. If an item is in doubt, mark with question mark and add a note of what to do to resolve the question. By: Dates:

VERY IMPORTANT:
MAKE SEPARATE LISTS OF SUBFUNCTIONS, SITE FEATURES, ROOMS AND SPACES, AND REVIEW THOSE SPACE NEEDS IN ACCORDANCE WITH THE SURVEY FORMS AT THE END OF THIS PREDESIGN DIVISION.

___ Secondary building functions.

___ Estimated construction budget.

___ Estimated construction deadline(s).

___ Estimated occupant population type(s) and sizes to fulfill stated function(s).

___ Special equipment to fulfill stated function(s).

___ Special furnishings to fulfill stated function(s).

___ Building or building division size(s) to accommodate population, circulation, furnishings and equipment.

___ Future building functions and populations.

USE SEPARATE SHEETS TO LIST OR RECORD SUBFUNCTIONS AND LISTS OF ROOMS OR SPATIAL FUNCTIONS.

___ Limits or allowable size of future expansion.

___ Existing facilities to be part of this project.

___ Existing facilities to use as design/planning examples for this project.

Add notes as needed after each checklisted task, such as:
Initials of who is to do it, when it's to start, when to review, who to coordinate with, and when it's to be finished.

Phase 1: Predesign 22
Programming and Predesign

EXTERNAL RESTRAINTS ON BUILDING AREA, SHAPE AND HEIGHT

Checkmark each item to be done and cross out the check when completed. Mark with a -- if an item is not to be done. If an item is in doubt, mark with question mark and add a note of what to do to resolve the question. By: Dates:

Legal restraints are checklisted in detail in the section on PERMITS AND APPROVALS in the next division, PHASE 2: SITE ANALYSIS.

___ Total lot dimensions and area.

___ Usable lot area.

___ Setback restrictions.

___ Other zoning restrictions.

___ Deed covenants.

___ Easements.

___ Rights of way.

___ Air rights.

___ Facade easement.

___ Existing construction.

___ Solar orientation.

___ Building shadow restrictions.

___ Required public spaces.

___ Groupings of population or function that require large open spaces.

___ Groupings of population or function that require courts or atriums.

___ Groupings of population or function that require direct access to exterior ground level.

___ Functions that require high ceiling interior spaces.

___ Functions requiring daylight.

___ Views.

___ Other.

Add notes as needed after each checklisted task, such as:
Initials of who is to do it, when it's to start, when to review, who to coordinate with, and when it's to be finished.

Phase 1: Predesign 23
Programming and Predesign

PRELIMINARY STRUCTURAL DECISIONS

Checkmark each item to be done and cross out the check when completed. Mark with a -- if an item is not to be done.
If an item is in doubt, mark with question mark and add a note of what to do to resolve the question. By: Dates:

___ Special span requirements related to space sizes, heights, and groupings.

___ Bay sizes.

___ Special soil conditions that restrict structural design.

___ Other special site conditions that restrict structure or construction:

    ___ Building in air space

    ___ Connection to adjacent structures

___ Anticipated building configuration:

    ___ Below ground

    ___ Ground level

    ___ Above ground

___ Wings/major divisions in building plan and configuration.

___ Core size, shape, and location.

___ Construction phases for structural work.

___ Anticipated structural frame.

___ Anticipated construction class/system.

___ Anticipated substructure systems and interior framing.

VERTICAL TRANSPORTATION OPTIONS

Checkmark each item to be done and cross out the check when completed. Mark with a -- if an item is not to be done.
If an item is in doubt, mark with question mark and add a note of what to do to resolve the question. By: Dates:

___ Elevator core:

    ___ Central/offset/detached

    ___ Interior/exterior

___ Freight elevators/special lifts.

___ Escalators.

Add notes as needed after each checklisted task, such as:
Initials of who is to do it, when it's to start, when to review, who to coordinate with, and when it's to be finished.

Phase 1: Predesign 24
Programming and Predesign

FIRE CODE REQUIREMENTS

Checkmark each item to be done and cross out the check when completed. Mark with a -- if an item is not to be done.
If an item is in doubt, mark with question mark and add a note of what to do to resolve the question. By: Dates:

A more detailed checklist of considerations and code data to obtain is included in the section on
BUILDING CODE AND FIRE CODE SEARCH at the end of PHASE 3: SCHEMATIC DESIGN.

___ Exit stairs/corridors.

___ Wall and partition ratings.

___ Fire barriers.

___ Door ratings.

INTERIOR PLANNING AND CONSTRUCTION

Checkmark each item to be done and cross out the check when completed. Mark with a -- if an item is not to be done.
If an item is in doubt, mark with question mark and add a note of what to do to resolve the question. By: Dates:

___ Anticipated Interior Partitions:

   ___ Framing

   ___ Finishes

   ___ Movable partitions, frames/finishes

   ___ Demountable partitions, frames/finishes

___ Anticipated ceiling construction:

   ___ Ceiling finishes

___ Anticipated primary space floor construction:

   ___ Primary space floor finishes

___ Anticipated secondary space floor construction:

   ___ Secondary space floor finishes

Add notes as needed after each checklisted task, such as:
Initials of who is to do it, when it's to start, when to review, who to coordinate with, and when it's to be finished.

Phase 1: Predesign 25
Programming and Predesign

EXTERIOR DESIGN AND CONSTRUCTION

Checkmark each item to be done and cross out the check when completed. Mark with a -- if an item is not to be done. If an item is in doubt, mark with question mark and add a note of what to do to resolve the question. By: Dates:

___ Special environmental conditions that restrict materials for the building envelope.

___ Anticipated exterior framing.

___ Anticipated exterior cladding type and material.

___ Anticipated exterior finishes.

___ Anticipated fenestration.

___ Fireproofing.

___ Weather protection.

___ Anticipated roof framing.

___ Anticipated finish roofing.

MECHANICAL

Checkmark each item to be done and cross out the check when completed. Mark with a -- if an item is not to be done. If an item is in doubt, mark with question mark and add a note of what to do to resolve the question. By: Dates:

___ HVAC system:

    ___ Perimeter

    ___ Interior

___ Solar components

___ Mechanical spaces:

    ___ Below grade

    ___ Interim level

    ___ Rooftop

Add notes as needed after each checklisted task, such as:
Initials of who is to do it, when it's to start, when to review, who to coordinate with, and when it's to be finished.

Phase 1: Predesign 26
Programming and Predesign

LIGHTING AND ELECTRICAL

Checkmark each item to be done and cross out the check when completed. Mark with a -- if an item is not to be done. If an item is in doubt, mark with question mark and add a note of what to do to resolve the question. By: Dates:

___ Anticipated lighting for primary space.

___ Anticipated lighting for secondary spaces.

___ Special power requirements.

___ Other.

USER QUESTIONNAIRE

This questionnaire is for establishing user desires and needs on a room-by-room basis and for identifying important relationships between rooms. Copy this form or your own version of it in quantity, and be sure every interior and exterior space is recorded.

PROJECT: SPACE NAME:
FLOOR OR LEVEL: DEPARTMENT:
SPACE SIZE: SPACE AREA: HEIGHT:

EXISTING EQUIVALENT ROOM NAME: NUMBER:
SIZE: AREA: HEIGHT:

TITLE OF USER(S):

NOTE OTHER SPACES THAT RELATE TO THIS SPACE AND RANK THE IMPORTANCE OF THEIR RELATIONSHIPS TO THIS SPACE:
Mandatory--4, Important--3, Desirable--1, Undesirable--X.

_____ ___   _____ ___
_____ ___   _____ ___
_____ ___   _____ ___
_____ ___   _____ ___
_____ ___   _____ ___

FURNITURE: SCHEMATIC FURNITURE PLAN:
(List by quantity, type, size, and code. Attach   (Note by type and #. Use bottom of second
separate list for complex room types.)   page or back of sheet for complex room types.
                                                   Note scale.)

_____
_____
_____
_____
_____

Add notes as needed after each checklisted task, such as:
Initials of who is to do it, when it's to start, when to review, who to coordinate with, and when it's to be finished.

Phase 1: Predesign 27
Programming and Predesign

USER QUESTIONNAIRE continued

This questionnaire is for establishing user desires and needs on a room-by-room basis and for identifying important relationships between rooms. Copy this form or your own version of it in quantity, and be sure every interior and exterior space is recorded.

EQUIPMENT:
(List by quantity, size and #.)
_____
_____
_____
_____

NUMBER OF PEOPLE:
(List by type/titles.)
_____
_____
_____
_____

ARCHITECTURAL FINISHES:
(Use code or abbreviations.)

FLOOR _____   WALLS _____

BASE _____   WAINSCOT _____

SPECIAL WALL _____   GLAZED WALL _____

DRAPES/BLINDS _____   CEILING _____

DOOR(S) _____   WINDOW(S) _____

ACOUSTICAL TREATMENT _____

STORAGE _____

CABINET WORK _____

PLUMBING FIXTURES   ELECTRICAL

_____   Outlets _____
_____   Switching _____
_____   Special loads _____

COMMUNICATIONS   LIGHTING

Phone _____   Ceiling _____
Computer _____   Spot _____
Dictaphone _____   Display _____
Pneumatic tube _____   Task _____
Delivery slot/window _____   Other _____
Other _____

HVAC
_____
_____

Add notes as needed after each checklisted task, such as:
Initials of who is to do it, when it's to start, when to review, who to coordinate with, and when it's to be finished.

Phase 2: Site Analysis 1
Predesign and Schematic Site Review

PROJECT MANAGEMENT CHECKLIST

Project Name/No: Notes by:

Dates Checked:

PLANNING AND ADMINISTRATION

Checkmark each item to be done and cross out the check when completed. Mark with a -- if an item is not to be done. If an item is in doubt, mark with question mark and add a note of what to do to resolve the question. By: Dates:

___ Confirm the accuracy of major features of the land survey by observation and measurement.

___ Obtain on-site photographs showing major site features. Use a site plan diagram with symbols and key numbers showing the camera positions and point of view for each photograph.

___ Use on-site photos later when appropriate to clarify schematic, design development, and working drawings.

___ Obtain cyclorama spliced photos of the site surroundings and views.

___ If the original site survey is not at an appropriate scale, have it photo reduced or enlarged. Obtain a screened shadow print transparency to use as a base sheet for site planning.

___ Obtain aerial or satellite photos of the site and its surroundings. Have an overall site photo produced at a scale to match the final site survey.

___ Obtain seasonal climate and microclimate statistics from the weather service.

___ Obtain or compute seasonal solar orientation data.

___ Confirm that the surrounding environment has been examined for negative factors that must be dealt with in the design. (See the FEASIBILITY AND FINANCIAL ANALYSIS section of PHASE 1 -- PREDESIGN.)

    ___ Noise sources.
    ___ Reflected sunlight and glare.
    ___ Odors.
    ___ Smoke.
    ___ Wind currents.
    ___ Periodic or chronic traffic congestion.
    ___ Shadows on the site.
    ___ Drainage onto the site.
    ___ Neighboring visual clutter.
    ___ Overhead utility lines.
    ___ Underground utilities.
    ___ Easements for future utilities.

Add notes as needed after each checklisted task, such as:
Initials of who is to do it, when it's to start, when to review, who to coordinate with, and when it's to be finished.

Phase 2: Site Analysis 2
Predesign and Schematic Site Review

PLANNING AND ADMINISTRATION continued

Checkmark each item to be done and cross out the check when completed. Mark with a -- if an item is not to be done.
If an item is in doubt, mark with question mark and add a note of what to do to resolve the question.   By: Dates:

___ Ask the consultants what site test data they need for their work, and what testing companies, laboratories, etc., they recommend.

___ Assist the client in obtaining necessary soil and related site tests and investigations.

___ Have design and consulting staff visit and examine the site.

___ Confirm that all consultants have the site data they need and that they have reviewed the data.

Add notes as needed after each checklisted task, such as:
Initials of who is to do it, when it's to start, when to review, who to coordinate with, and when it's to be finished.

Phase 2: Site Analysis 3
Environmental Impact Report

PROJECT MANAGEMENT CHECKLIST

Project Name/No: Notes by:

Dates Checked:

Environmental Impact Reports or Statements as required by the client or governmental agencies should be planned and checked to ensure that all relevant items in this checklist are included.

PLANNING AND ADMINISTRATION

Checkmark each item to be done and cross out the check when completed. Mark with a -- if an item is not to be done. If an item is in doubt, mark with question mark and add a note of what to do to resolve the question. By: Dates:

___ Decide extent of content.

    ___ Obtain legal requirements of format and content.
    ___ Obtain public agency checklists.
    ___ Obtain samples of similar reports in public records.
    ___

___ Decide the structure and focus of the report.

    ___ General and varied environmental issues/Non-focused.
    ___ Focused/Issue Oriented.
    ___

___ Decide and check details of the report on project evaluation.

    ___ Aesthetic enhancement.
    ___ Neighborhood or local upgrading.
    ___ Enhancement of neighborhood or local economy.
    ___ Land use improvements.
    ___ Upgrading of neighboring or local properties.
    ___ Traffic flow and parking improvement.
    ___

    ___ Air quality protection or improvement.
    ___ Microclimate, air motion and humidity improvement.
    ___ Water quality protection or improvement.
    ___ Improved surface water flow.
    ___ Improved ground water flow.
    ___ Earth slide and erosion prevention.
    ___

Add notes as needed after each checklisted task, such as:
Initials of who is to do it, when it's to start, when to review, who to coordinate with, and when it's to be finished.

Phase 2: Site Analysis 4
Environmental Impact Report

PLANNING AND ADMINISTRATION continued

Checkmark each item to be done and cross out the check when completed. Mark with a -- if an item is not to be done. If an item is in doubt, mark with question mark and add a note of what to do to resolve the question.   By: Dates:

    ___ Animal life preservation or enhancement.
    ___ Plant life protection or enhancement.
    ___ Historical preservation.
    ___ Archaeological protection.
    ___ Safety enhancement.
    ___ Noise abatement.
    ___ Glare and reflectance prevention.
    ___

    ___ Natural resources development.
    ___ Improved market for local services.
    ___ Tax revenue increase.
    ___ Improved market for utility services.
    ___ Neighborhood or local security improvement.
    ___ Health and recreation enhancement.
    ___ Local ethnic values recognition.
    ___

___ Decide and check environmental impact considerations during project planning and construction phases.

    ___ Acquisition of property.
    ___ Relocation of tenants or owners.
    ___ Planning. (Effect of anticipation of the project on other property owners, real estate speculators, etc.)
    ___ Demolition of existing structures/land clearing.
    ___ Construction.
    ___ Operation of the facility.
    ___ Future related or contiguous development.
    ___

Add notes as needed after each checklisted task, such as:
Initials of who is to do it, when it's to start, when to review, who to coordinate with, and when it's to be finished.

Phase 2: Site Analysis 5
Permits and Approvals

PROJECT MANAGEMENT CHECKLIST

Project Name/No: Notes by:

Dates Checked:

Most of the research and administration tasks listed below will normally be done by only one or two persons. In such cases you can skip the delegation and deadline date coding.

Also see the BUILDING CODE AND FIRE CODE SEARCH section of PHASE 3: SCHEMATIC DESIGN.

RESEARCH AND ADMINISTRATION

Checkmark each item to be done and cross out the check when completed. Mark with a -- if an item is not to be done. If an item is in doubt, mark with question mark and add a note of what to do to resolve the question.   By: Dates:

___ Create a list of governing agencies, agency representatives, and all applicable codes, regulations, and ordinances that pertain to the project. Note specific codes and ordinance numbers. (See the forms that follow in this section of the manual.)

___ Make a Directory of agency names, representative names, addresses, and phone numbers.

___ Locate revisions of codes and ordinances published after the main body of the code or ordinance volume.

___ Identify as yet unpublished revisions of codes and ordinances.

___ Identify ambiguities, contradictions, duplications, or overlaps in written codes and regulatory agency jurisdictions.

___ Write memos to all parties asking for clarification of contradictory or ambiguous regulations.

___ Identify sequential requirements such as approvals required prior to obtaining other approvals.

___ Make a calendar and checklist of all permit and approval processes.

___ Identify and assign staff and management responsible for research, agency contact, public meeting representation, etc.

___ Request information on any special code requirements known to the client.

___ Create a worksheet for each regulatory agency concerned with the project. (See the worksheet form on the next page.)

Add notes as needed after each checklisted task, such as:
Initials of who is to do it, when it's to start, when to review, who to coordinate with, and when it's to be finished.

Phase 2: Site Analysis 6
Permits and Approvals

CODES AND REGULATIONS RECORD SHEET

Project Name & Number:

Project Location/Address:

Client Name, Address & Phone:

Project Manager:
Principal in Charge:

Department, Agency, or Utility:
Address:

Agency Representative:
Phone and Extension:

Representative's Superior:
Phone and Extension:

Representative's Assistant:
Phone and Extension:

Applicable Code, Ordinance or Rule Numbers & Revisions:

Coordinate With Other Agencies:

Dates of Reviews: Topics: Results:

Add notes as needed after each checklisted task, such as:
Initials of who is to do it, when it's to start, when to review, who to coordinate with, and when it's to be finished.

Phase 2: Site Analysis 7
Permits and Approvals

REGULATORY AGENCY MASTER LIST

Checkmark each item to be done and cross out the check when completed. Mark with a -- if an item is not to be done. If an item is in doubt, mark with question mark and add a note of what to do to resolve the question. By: Dates:

AGENCY, CODE, AND SOURCE OF REGULATION DATA:

___ CITY OR REGIONAL BUILDING CODE (See separate checklist.)

___ CITY ZONING OFFICE (See separate checklist that follows.)

___ ENVIRONMENTAL AGENCY (See separate EIR CHECKLIST.)

___ REGIONAL PLANNING COMMISSION

___ REDEVELOPMENT AGENCY

___ DESIGN CONTROL

___ SIGN CONTROL

___ HANDICAPPED

___ FIRE PREVENTION

___ HEALTH DEPARTMENT

___ INDUSTRIAL SAFETY

___ INDUSTRIAL WASTE

___ ROADS AND TRAFFIC

___ GRADING AND DRAINAGE CONTROL

___ UTILITIES:
    ___ Water
    ___ Electric Power
    ___ Telephone
    ___ Other communications
    ___ Cable
    ___ Sewer
    ___ Gas

SPECIAL USE PERMITS:

___ MEDICAL

___ EDUCATIONAL

___ RECREATIONAL

Add notes as needed after each checklisted task, such as:
Initials of who is to do it, when it's to start, when to review, who to coordinate with, and when it's to be finished.

Phase 2: Site Analysis 8
Permits and Approvals

ZONING

Checkmark each item to be done and cross out the check when completed. Mark with a -- if an item is not to be done. If an item is in doubt, mark with question mark and add a note of what to do to resolve the question. By: Dates:

ZONE STATUS:

___ Existing Site Zoning.
___ Zone Required.

APPROVALS REQUIRED:

___ Variance.
___ Exception.
___ Use Permit.

LEGAL ACTIONS:

___ Prime Agency.
___ Appeals Agency.
___ Court Jurisdiction.

ZONE SETBACK REQUIREMENTS:

___ Front.
___ Side.
___ Rear.
___ Arc and angle property lines.

ZONE HEIGHT LIMITS:

___ Building height-to-land area ratios.
___ Exception trades for dedicated public space.
___ Air rights.
___ View entitlements.
___ Sun and shade entitlements.

LOT SIZE AND BUILDABLE AREA:

___ Site area size.
___ Gross site area within setbacks.
___ Area of existing construction to remain.
___ Size of areas restricted from construction.
___ Net ground floor or air right buildable area.
___ Allowable below grade building area.
___ Allowable upper air projections or extensions.
___ Net upper story level buildable areas, floor by floor.

Add notes as needed after each checklisted task, such as:
Initials of who is to do it, when it's to start, when to review, who to coordinate with, and when it's to be finished.

Phase 2: Site Analysis 9
Permits and Approvals

ZONING continued

Checkmark each item to be done and cross out the check when completed. Mark with a -- if an item is not to be done. If an item is in doubt, mark with question mark and add a note of what to do to resolve the question. By: Dates:

DISTANCES TO ADJACENT CONSTRUCTION:

___ Front to front.
___ Front to side.
___ Front to rear.
___ Side to side.
___ Side to rear.
___ Rear to rear.
___ Wells and courtyards.

PARKING:

___ Required parking stall-to-occupancy ratios.
___ Number of parking stalls required.
___ Required parking stall sizes.
___ Number and size of allowable compact car stalls.
___ Required aisle sizes relative to parking angles:
    ___ 90 degree.
    ___ 60 degree.
    ___ 45 degree.
    ___ 30 degree.

___ Required wheel stops and curbs.
___ Driveways.
___ Parking driveway apron/ramp slope and width limits.

MISCELLANEOUS ZONING REQUIREMENTS:

___ Fencing--allowable types, heights, and distances from property and building lines.
___ Yard walls--allowable types, heights, and distances from property and building lines.
___ Exterior building display lighting.
___ Exterior yard lighting.
___ Building yard security.
___ Required landscaping or open areas.
___ Required pavement drainage.
___ Restrictions on trash collection area.
___ Zoning restrictions regarding handicap access.
___ Other.

Add notes as needed after each checklisted task, such as:
Initials of who is to do it, when it's to start, when to review, who to coordinate with, and when it's to be finished.

Phase 3: Schematic Design 1
Construction Cost Estimating

PROJECT MANAGEMENT CHECKLIST

Project Name/No: Notes by:

Dates Checked:

ADMINISTRATION -- SCHEMATICS THROUGH CONSTRUCTION DOCUMENTS

Checkmark each item to be done and cross out the check when completed. Mark with a -- if an item is not to be done.
If an item is in doubt, mark with question mark and add a note of what to do to resolve the question. By: Dates:

___ Confirm thoroughness of construction cost estimates (AIA Doc. B131, B231, B331).

    ___ Estimate of Probable Construction Cost.
    ___ Detailed Estimate of Probable Construction Cost as an Additional Service.
    ___ Life Cycle Costing.
    ___

___ Schedule times of project cost estimates and estimate updates.

    ___ Predesign and programming.
    ___ Concept package for financing.
    ___ Schematics.
    ___ Design Development.
    ___ Construction Documents at _____ % completion.
    ___ Construction Documents, prebid.
    ___

___ Decide type of construction cost estimating system to use and at what phases different systems may be used.

    ___ Quantity survey.
    ___ Square foot unit cost (AIA Doc. D101).
    ___ Volume unit cost (AIA Doc. D101).
    ___ In-Place unit cost.
    ___

___ Establish a construction estimate form for Schematic and Design Development phases, including cost data on:

    ___ Overall construction:
        ___ Foundation.
        ___ Structural framing.
        ___ Roofing.
        ___ Envelope.
        ___ Interiors.
        ___ Floors.
        ___ Walls.
        ___ Ceilings.
        ___ Cabinets.

Add notes as needed after each checklisted task, such as:
Initials of who is to do it, when it's to start, when to review, who to coordinate with, and when it's to be finished.

Phase 3: Schematic Design 2
Construction Cost Estimating

ADMINISTRATION -- SCHEMATICS THROUGH CONSTRUCTION DOCUMENTS continued
Checkmark each item to be done and cross out the check when completed. Mark with a -- if an item is not to be done.
If an item is in doubt, mark with question mark and add a note of what to do to resolve the question. By: Dates:

    ___ HVAC.
    ___ Plumbing.
    ___ Fire protection.
    ___ Electrical.
    ___ Vertical transportation.
    ___ Sitework.
    ___ Special equipment and furnishings.
    ___ Contingency allowance.
    ___

___ Decide building labor and materials cost information source(s).

    ___ In-house records.
    ___ F.W. Dodge.
    ___ R.S. Means.
    ___ Local or regional sources.
    ___ Computerized systems.
    ___

___ Decide on, or schedule a decision on the design of detailed construction estimate forms for Construction Document and Bidding phases.

Add notes as needed after each checklisted task, such as:
Initials of who is to do it, when it's to start, when to review, who to coordinate with, and when it's to be finished.

Phase 3: Schematic Design 3
Schematic Design and Documentation

PROJECT MANAGEMENT CHECKLIST

Project Name/No: Notes by:

Dates Checked:

     Starting with SCHEMATIC DESIGN AND DOCUMENTATION, this manual includes task group index identification numbers from the AIA'S SCOPE OF DESIGNATED SERVICES (DOC. B162).

DESIGN PROCESS

Checkmark each item to be done and cross out the check when completed. Mark with a -- if an item is not to be done. If an item is in doubt, mark with question mark and add a note of what to do to resolve the question. By: Dates:

___ Confirm and update the building program's stated functional, occupancy and spatial requirements with the client.

___ If the building's size, configuration, stacking, and structural system, as well as other major decisions, have not yet been made, follow the planning steps checklisted in the PREDESIGN AND PROGRAMMING section of PHASE 1: PREDESIGN.

___ Decide on or confirm a structural module with the engineer.

___ Decide on or confirm the interior partitioning and ceiling module.

___ Plan the disposition of major or departmental spaces as per the spatial diagrams generated from the program. Obtain preliminary approval from the users and/or client.

___ Plan the disposition of minor or subdepartmental spaces within the planned larger spaces. Get preliminary approvals from the users and/or client.

___ Prepare notes and diagrams as guides for consultants' preliminary work.

ADMINISTRATION -- UPDATES AFTER PREDESIGN AND PROGRAMMING

Checkmark each item to be done and cross out the check when completed. Mark with a -- if an item is not to be done. If an item is in doubt, mark with question mark and add a note of what to do to resolve the question. By: Dates:

___ Compare the latest predesign and programming requirements with the client's budget. Confirm the budget agreement or settle any contradictions between stated program needs and available funding.

___ Before finalizing new consultant contracts, review service and contract terms with the client and obtain written client approvals.

___ Obtain the client's written approval of the work of the predesign and programming phase.

___ Obtain the client's written agreement to proceed with the schematic design.

Add notes as needed after each checklisted task, such as:
Initials of who is to do it, when it's to start, when to review, who to coordinate with, and when it's to be finished.

Phase 3: Schematic Design 4
Schematic Design and Documentation

DISCIPLINES COORDINATION AND DOCUMENT CHECKING

Checkmark each item to be done and cross out the check when completed. Mark with a -- if an item is not to be done. If an item is in doubt, mark with question mark and add a note of what to do to resolve the question. By: Dates:

___ Require all consultants to do their schematics following the same scale, format, and drawing positioning as the architectural drawings. (Preferably, provide architectural base sheet schematics and require consultants to do their work on transparency overlays.)

___ Create or obtain lists of special building equipment and fixtures required by the client that may affect consultants' work, and distribute the lists to the appropriate consultants.

___ Review architectural schematic diagrams while in process with structural, mechanical, electrical, transportation, and other consultants. Conduct one or more group meetings to allow consultants to compare their work. (Preferably, work should be compared in layers with overlay transparencies.)

___ Reach agreement on specific appropriate structural, construction, mechanical, and other building systems.

___ Confirm that the selected engineering and construction systems are compatible with one another.

___ Obtain estimates of spatial requirements for appurtenances and engineered systems.

___ Confirm that the prime building designer is fully informed and updated regarding all consultants' information.

___ Coordinate engineering schematic building diagrams.

AGENCY CONSULTING, REVIEW, AND APPROVALS

This section is covered by the PERMITS AND APPROVALS section of PHASE 2: SITE ANALYSIS division, and the BUILDING CODE AND FIRE CODE SEARCH section of PHASE 3: SCHEMATIC DESIGN.

Add notes as needed after each checklisted task, such as:
Initials of who is to do it, when it's to start, when to review, who to coordinate with, and when it's to be finished.

Phase 3: Schematic Design 5
Schematic Design and Documentation

OWNER-SUPPLIED DATA COORDINATION

Checkmark each item to be done and cross out the check when completed. Mark with a -- if an item is not to be done.
If an item is in doubt, mark with question mark and add a note of what to do to resolve the question. By: Dates:

___ Reconfirm the program's functional, occupancy, and spatial requirements with the client.

___ Compare the developed design with the client's budget. Confirm the budget agreement or settle any contradictions between stated program needs and available funding.

___ Reconfirm the client's written approval of the building program.

___ Reconfirm the client's written agreement to proceed with the Schematic Design phase.

___ Obtain a list of the client's special building equipment and fixtures requirements that may affect the consultants' work.

ARCHITECTURAL DESIGN AND DOCUMENTATION

Checkmark each item to be done and cross out the check when completed. Mark with a -- if an item is not to be done.
If an item is in doubt, mark with question mark and add a note of what to do to resolve the question. By: Dates:

___ Confirm the sources and accuracy of the budget, program, and legal requirements.

___ If there are changes in the design staff between the Predesign/Programming phase and Schematic Design, confirm that new staff have acquired and assimilated all previous design data.

___ Schedule the architectural Schematic Design drawings:
    ___ Site Plan.
    ___ Floor Plans.
    ___ Roof Plan.
    ___ Cross Sections.
    ___ Exterior Elevations.
    ___ Interior Elevations.
    ___ Wall Sections.
    ___ Design Details.

___ Compare schematic plans, sections and elevations with the program.

___ Review schematic plans, sections, and elevations for rough construction cost estimates and compare with the budget.

___ Compare schematic plans, sections and elevations with code and regulatory requirements.

Add notes as needed after each checklisted task, such as:
Initials of who is to do it, when it's to start, when to review, who to coordinate with, and when it's to be finished.

Phase 3: Schematic Design 6
Schematic Design and Documentation

STRUCTURAL DESIGN AND DOCUMENTATION

Checkmark each item to be done and cross out the check when completed. Mark with a -- if an item is not to be done. If an item is in doubt, mark with question mark and add a note of what to do to resolve the question. By: Dates:

Also see the PRELIMINARY STRUCTURAL DECISIONS checklist, in the PREDESIGN AND PROGRAMMING section of PHASE 1: PREDESIGN.

___ Schedule structural and architectural coordination meetings.

___ Schedule structural, mechanical, and architectural drawing cross-checking meetings.

___ Review and reach agreement with the structural engineer on the number and content of structural Schematic Design documents:
  ___ Design criteria.
  ___ Structural grid or system.
  ___ Alternative grids or structural systems.
  ___ Schematic framing plans and sections.
  ___ Schematic foundation plan.
  ___ Schematic structural cross section(s).
  ___ Calculations.
  ___ Required clearances for other construction.

___ Schedule completion dates for structural schematic drawings.

___ Confirm with the structural engineer that the proposed structural systems satisfy all legal requirements.

___ Obtain preliminary estimates for probable structural systems construction costs.

MECHANICAL DESIGN AND DOCUMENTATION

Checkmark each item to be done and cross out the check when completed. Mark with a -- if an item is not to be done. If an item is in doubt, mark with question mark and add a note of what to do to resolve the question. By: Dates:

___ Schedule structural, mechanical, and architectural drawing cross-checking meetings.

___ Review and reach agreement with the mechanical engineer on the number and content of mechanical Schematic Design data.

  ___ Design criteria:
    ___ Energy use and conservation.
    ___ HVAC system type and standard.
    ___ Plumbing supply and drain types and standards.
    ___ Fire protection systems.
    ___ Mechanical equipment estimated spatial requirements in plan.
    ___ Mechanical equipment estimated spatial requirements in section.
    ___ Alternate mechanical systems.

  ___ Preliminary equipment and materials schedules.
  ___ Outline specifications.

Add notes as needed after each checklisted task, such as:
Initials of who is to do it, when it's to start, when to review, who to coordinate with, and when it's to be finished.

Phase 3: Schematic Design 7
Schematic Design and Documentation

MECHANICAL DESIGN AND DOCUMENTATION

Checkmark each item to be done and cross out the check when completed. Mark with a -- if an item is not to be done. If an item is in doubt, mark with question mark and add a note of what to do to resolve the question. By: Dates:

___ Schedule completion dates for interim and final mechanical schematic drawings.

___ Obtain preliminary estimates for probable mechanical systems construction costs.

ELECTRICAL DESIGN AND DOCUMENTATION

Checkmark each item to be done and cross out the check when completed. Mark with a -- if an item is not to be done. If an item is in doubt, mark with question mark and add a note of what to do to resolve the question. By: Dates:

___ Schedule electrical and architectural coordination meetings.

___ Review and reach agreement with the electrical engineer on the number and content of electrical schematic documents.

    ___ Schematic plans, sections, and notes to show:
        ___ Reflected ceiling lighting plans.
        ___ Power and switching.
        ___ Communications equipment, chases, and outlets.
        ___ Fire protection and alarms.
        ___ Security systems.
        ___ Major electrical equipment sizes and locations.
        ___ Electrical vaults, transformer rooms.
        ___ Estimated spatial requirements for equipment and service.
        ___ Alternate systems.

___ Schedule completion dates for interim and final electrical Schematic Design documents.

___ Check and confirm compliance of the proposed building's electrical system design with codes and utility company requirements.

___ Obtain preliminary estimates for probable electrical systems construction costs.

Add notes as needed after each checklisted task, such as:
Initials of who is to do it, when it's to start, when to review, who to coordinate with, and when it's to be finished.

Phase 3: Schematic Design 8
Schematic Design and Documentation

CIVIL DESIGN AND DOCUMENTATION

Checkmark each item to be done and cross out the check when completed. Mark with a -- if an item is not to be done. If an item is in doubt, mark with question mark and add a note of what to do to resolve the question. By: Dates:

___ Confirm that results of previously requested site tests have been received and transmitted to the client, consultants, and the design team.

___ Identify additional tests that may be required, and update the Test Log and file.

___ Schedule civil, structural, landscaping, and architectural coordination and drawing cross-checking meetings.

___ Review and reach agreement with the civil engineer on the number and content of civil engineering Schematic Design documents.

    ___ Site plan diagrams and notes to show:

        ___ Utility systems, on site and off.
        ___ Fire protection systems.
        ___ Cut and fill.
        ___ Excavations.
        ___ Irrigation.
        ___ Drainage.
        ___ Alternate systems of civil work.
        ___

___ Schedule completion dates for interim and final civil Schematic documents.

___ Check and confirm compliance of sitework and civil engineering design with codes and regulations.

___ Prepare preliminary estimates for probable civil engineering related construction costs.

Add notes as needed after each checklisted task, such as:
Initials of who is to do it, when it's to start, when to review, who to coordinate with, and when it's to be finished.

Phase 3: Schematic Design 9
Schematic Design and Documentation

LANDSCAPE DESIGN AND DOCUMENTATION

Checkmark each item to be done and cross out the check when completed. Mark with a -- if an item is not to be done. If an item is in doubt, mark with question mark and add a note of what to do to resolve the question. By: Dates:

___ Schedule landscaping, architectural, and multidiscipline coordination meetings and/or procedures.

___ Review and reach agreement with the landscape architect on the number and content of landscape design Schematic Design drawings and notes.

    ___ Design criteria.
    ___ Preliminary planting and landscaping planning.
    ___ Site-related plumbing work.
    ___ Site-related electrical work.
    ___ Alternate landscaping concepts.
    ___

___ Schedule completion dates for interim and final landscape Schematic Design documents.

___ Obtain preliminary estimates for probable landscaping development costs.

INTERIOR DESIGN AND DOCUMENTATION

Checkmark each item to be done and cross out the check when completed. Mark with a -- if an item is not to be done. If an item is in doubt, mark with question mark and add a note of what to do to resolve the question. By: Dates:

___ Schedule interior design and architectural coordination meetings.

___ Review and reach agreement with the interior designer on the number and content of interior Schematic Design documents.

    ___ Preliminary interior partition landscaping.
    ___ Preliminary furniture planning.
    ___ Materials and finishes palette.

___ Schedule completion dates for interim and final interior Schematic Design documents.

___ Obtain preliminary estimates for probable cost of interior design partitions and furnishings.

Add notes as needed after each checklisted task, such as:
Initials of who is to do it, when it's to start, when to review, who to coordinate with, and when it's to be finished.

Phase 3: Schematic Design 10
Schematic Design and Documentation

MATERIALS RESEARCH AND SPECIFICATIONS

Checkmark each item to be done and cross out the check when completed. Mark with a -- if an item is not to be done. If an item is in doubt, mark with question mark and add a note of what to do to resolve the question. By: Dates:

For a detailed specifications checklist, see the section on SPECIFICATION WRITING AND COORDINATION in PHASE 5: CONSTRUCTION DOCUMENTS.

___ Start research on materials, equipment, fixtures, and building systems. Create a products and materials file or binder.

___ Start a list of first choices and alternative choices in materials, finishes, etc.

___ Start project outline specifications or contents list in coordination with schematic architectural drawings.

___ Review with consultants the extent of outline specifications required from them at this phase.

PROJECT DEVELOPMENT SCHEDULING

Checkmark each item to be done and cross out the check when completed. Mark with a -- if an item is not to be done. If an item is in doubt, mark with question mark and add a note of what to do to resolve the question. By: Dates:

___ Create or update the job calendar of estimated phase starts and completions.
    ___ Schematic Design.
    ___ Design Development.
    ___ Construction Documents.
    ___ Bidding/Negotiation.
    ___ Contract Administration.
    ___ Postconstruction.

___ Distribute copies of the new or updated job calendar to all job participants.

___ Create a schedule for budget and progress reviews (AIA Doc. F721, 723, F800 Series).

Add notes as needed after each checklisted task, such as:
Initials of who is to do it, when it's to start, when to review, who to coordinate with, and when it's to be finished.

Phase 3: Schematic Design 11
Schematic Design and Documentation

CONSTRUCTION COST ESTIMATING

Checkmark each item to be done and cross out the check when completed. Mark with a -- if an item is not to be done. If an item is in doubt, mark with question mark and add a note of what to do to resolve the question. By: Dates:

For a detailed cost estimating checklist, see the section on CONSTRUCTION COST ESTIMATING.

___ Have all consultants prepare construction cost estimates for their phases of work.

___ Prepare an overall construction cost estimate with a clearly stated contingency factor.

PRESENTATION

Checkmark each item to be done and cross out the check when completed. Mark with a -- if an item is not to be done. If an item is in doubt, mark with question mark and add a note of what to do to resolve the question. By: Dates:

See the detailed presentation media checklist in the MARKETING AND PRESENTATION MANAGEMENT section of PHASE 1: PREDESIGN.

___ Plan appropriate presentation media: models, renderings, slides, video, computer graphics. Identify client preferences.

___ Prepare presentations of optional design features and variations to be compared and decided upon by the client.

___ Prepare building floor area calculations, building volume, usable area ratios, and other numerical comparisons with program requirements.

___ Prepare presentation data on preferred construction methods.

___ Prepare data on costs and availability of special equipment and furnishings.

___ Have all consultants prepare construction cost estimates for their phases of work.

___ Prepare an overall construction cost estimate with a clearly noted contingency factor.

___ Prepare the presentation materials in such a form as to maximize possible reuse in design development. Keep solid-line drawing on base sheets, for example, and do rendered entourage or design variations on overlays in order to keep original solid line work available for reuse.

___ Present the Schematic Design and cost data.

Add notes as needed after each checklisted task, such as:
Initials of who is to do it, when it's to start, when to review, who to coordinate with, and when it's to be finished.

Phase 3: Schematic Design 12
Schematic Design and Documentation

POSTPRESENTATION

Checkmark each item to be done and cross out the check when completed. Mark with a -- if an item is not to be done. If an item is in doubt, mark with question mark and add a note of what to do to resolve the question.   By: Dates:

___ Identify changes in the Schematic Design required by the client.

___ Note any extended repercussions from design changes and review with the client any revisions of the Scope of Work.

___ Identify any contradictions between requested design changes and the original design program or prior client/designer decisions. Review these with the client.

___ Identify any contradictions between requested design changes and the original design program or prior client/designer decisions. Review these with the client.

___ Obtain the client's written agreement to proceed with the next phase of design development.

___ Establish a schedule and completion dates for review of upcoming work in PHASE 4: DESIGN DEVELOPMENT.

___ Prepare and submit billing during this phase of work as per the design service contract.

Add notes as needed after each checklisted task, such as:
Initials of who is to do it, when it's to start, when to review, who to coordinate with, and when it's to be finished.

Phase 3: Schematic Design 13
Building Code & Fire Code Search

PROJECT MANAGEMENT CHECKLIST

Project Name/No: Notes by:

Dates Checked:

    Also see the PERMITS AND APPROVALS section in PHASE 2: SITE ANALYSIS.

Project Location/Address:

Governing Code(s) and Release Date(s):

Code, Revised Edition(s):

Agency Representative(s):

BUILDING CODE SEARCH

Checkmark each item to be done and cross out the check when completed. Mark with a -- if an item is not to be done. If an item is in doubt, mark with question mark and add a note of what to do to resolve the question. By: Dates:

___ Identify and note all governing building code requirements as listed on the pages that follow.

___ FIRE ZONE
    Code sections and pages: _____
    Check: ___ Predesign ___ Schematics ___ Design Development

___ OCCUPANCY TYPE
    Code sections and pages: _____
    Check: ___ Predesign ___ Schematics ___ Design Development

___ MIXED OCCUPANCY
    Code sections and pages: _____
    Check: ___ Predesign ___ Schematics ___ Design Development

___ ALLOWABLE CONSTRUCTION TYPE
    Code sections and pages: _____
    Check: ___ Predesign ___ Schematics ___ Design Development

___ LOCATION ON PROPERTY
    Code sections and pages: _____
    Check: ___ Predesign ___ Schematics ___ Design Development

Add notes as needed after each checklisted task, such as:
Initials of who is to do it, when it's to start, when to review, who to coordinate with, and when it's to be finished.

Phase 3: Schematic Design 14
Building Code & Fire Code Search

BUILDING CODE SEARCH continued

Checkmark each item to be done and cross out the check when completed. Mark with a -- if an item is not to be done. If an item is in doubt, mark with question mark and add a note of what to do to resolve the question. By: Dates:

___ MULTIPLE BUILDINGS ON THE SAME SITE
  Code sections and pages: _____
  Check: ___ Predesign ___ Schematics ___ Design Development

___ ALLOWABLE GROUND FLOOR AREA
  Code sections and pages: _____
  Check: ___ Predesign ___ Schematics ___ Design Development

___ ALLOWABLE FLOOR AREA--MULTISTORY BUILDING
  Code sections and pages: _____
  Check: ___ Predesign ___ Schematics ___ Design Development

___ ALLOWABLE HEIGHT AND NUMBER OF STORIES
  Code sections and pages: _____
  Check: ___ Predesign ___ Schematics ___ Design Development

___ ALLOWABLE OCCUPANT LOAD
  Code sections and pages: _____
  Check: ___ Predesign ___ Schematics ___ Design Development

___ EXIT REQUIREMENTS
  Code sections and pages: _____
  Check: ___ Predesign ___ Schematics ___ Design Development

___ STRUCTURAL AND CONSTRUCTION LIMITATIONS
  Code sections and pages: _____
  Check: ___ Predesign ___ Schematics ___ Design Development

___ MECHANICAL SYSTEM LIMITATIONS
  Code sections and pages: _____
  Check: ___ Predesign ___ Schematics ___ Design Development

___ ENERGY CONSERVATION
  Code sections and pages: _____
  Check: ___ Predesign ___ Schematics ___ Design Development

Add notes as needed after each checklisted task, such as:
Initials of who is to do it, when it's to start, when to review, who to coordinate with, and when it's to be finished.

Phase 3: Schematic Design 15
Building Code & Fire Code Search

FIRE CODE SEARCH

Checkmark each item to be done and cross out the check when completed. Mark with a -- if an item is not to be done. If an item is in doubt, mark with question mark and add a note of what to do to resolve the question. By: Dates:

___ Identify and note all governing fire code requirements as listed on the pages that follow.

___ AUTOMATIC SPRINKLERS (type and coverage)
    Check: ___ Broadscope drawings ___ Details ___ Specifications
    Coord: ___ Plumbing

___ SMOKE DETECTORS (type and locations)
    Check: ___ Broadscope drawings ___ Details ___ Specifications
    Coord: ___ Electrical

___ FIRE EXTINGUISHERS (types and locations)
    Check: ___ Broadscope drawings ___ Details ___ Specifications

___ FIRE HOSE CABINETS (type and locations)
    Check: ___ Broadscope drawings ___ Details ___ Specifications
    Coord: ___ Plumbing

___ DRY STANDPIPES (sizes and locations)
    Check: ___ Broadscope drawings ___ Details ___ Specifications
    Coord: ___ Plumbing

___ WET STANDPIPES (sizes and locations)
    Check: ___ Broadscope drawings ___ Details ___ Specifications
    Coord: ___ Plumbing

___ COMBINATION STANDPIPES (sizes and locations)
    Check: ___ Broadscope drawings ___ Details ___ Specifications
    Coord: ___ Plumbing

___ FIRE ALARMS (types and locations)
    Check: ___ Broadscope drawings ___ Details ___ Specifications
    Coord: ___ Electrical

___ AUTOMATIC CORRIDOR CLOSERS (types and locations)
    Check: ___ Broadscope drawings ___ Details ___ Specifications

Add notes as needed after each checklisted task, such as:
Initials of who is to do it, when it's to start, when to review, who to coordinate with, and when it's to be finished.

Phase 3: Schematic Design 16
Building Code & Fire Code Search

FIRE CODE SEARCH continued

Checkmark each item to be done and cross out the check when completed. Mark with a -- if an item is not to be done. If an item is in doubt, mark with question mark and add a note of what to do to resolve the question.   By: Dates:

__ ABOVE-CEILING FIRE BARRIERS (types and locations)
   Check: __ Broadscope drawings __ Details __ Specifications

__ ATTIC FIRE BARRIERS (types and locations)
   Check: __ Broadscope drawings __ Details __ Specifications

__ LIGHTED EXIT SIGNS
   Check: __ Broadscope drawings __ Details __ Specifications
   Coord: __ Electrical

__ BATTERY-POWERED EMERGENCY LIGHTING
   Check: __ Broadscope drawings __ Details __ Specifications
   Coord: __ Electrical

__ SMOKE VENTILATION
   Check: __ Broadscope drawings __ Details __ Specifications
   Coord: __ Mechanical

__ HEAT VENTILATION
   Check: __ Broadscope drawings __ Details __ Specifications
   Coord: __ Mechanical

__ ELEVATOR FIRE RESPONSE
   Check: __ Specifications
   Coord: __ Electrical

Add notes as needed after each checklisted task, such as:
Initials of who is to do it, when it's to start, when to review, who to coordinate with, and when it's to be finished.

Phase 3: Schematic Design 17
Building Code & Fire Code Search

FIRE CODE REQUIREMENTS -- FIRE-RATED CONSTRUCTION

Checkmark each item to be done and cross out the check when completed. Mark with a -- if an item is not to be done.
If an item is in doubt, mark with question mark and add a note of what to do to resolve the question. By: Dates:

___ EXTERIOR BEARING WALLS
    Check: ___ Broadscope drawings ___ Details ___ Specifications
    Coord: ___ Structural

___ EXTERIOR NON-BEARING WALLS
    Check: ___ Broadscope drawings ___ Details ___ Specifications
    Coord: ___ Structural

___ EXTERIOR DOORS
    Check: ___ Broadscope drawings ___ Details ___ Specifications

___ FENESTRATION
    Check: ___ Broadscope drawings ___ Details ___ Specifications

___ INTERIOR DOORS
    Check: ___ Broadscope drawings ___ Details ___ Specifications

___ INTERIOR BEARING WALLS
    Check: ___ Broadscope drawings ___ Details ___ Specifications
    Coord: ___ Structural

___ INTERIOR NON-BEARING PARTITIONS
    Check: ___ Broadscope drawings ___ Details ___ Specifications

___ INTERIOR MOVABLE PARTITIONS
    Check: ___ Broadscope drawings ___ Details ___ Specifications

___ INTERIOR DEMOUNTABLE PARTITIONS
    Check: ___ Broadscope drawings ___ Details ___ Specifications

___ INTERIOR LOW-HEIGHT PARTITIONS
    Check: ___ Broadscope drawings ___ Details ___ Specifications

___ INTERIOR PERMANENT PARTITIONS
    Check: ___ Broadscope drawings ___ Details ___ Specifications
    Coord: ___ Structural

___ INTERIOR PARTITION FINISHES
    Check: ___ Broadscope drawings ___ Details ___ Specifications

___ SHAFT WALLS
    Check: ___ Broadscope drawings ___ Details ___ Specifications
    Coord: ___ Structural

Add notes as needed after each checklisted task, such as:
Initials of who is to do it, when it's to start, when to review, who to coordinate with, and when it's to be finished.

Phase 3: Schematic Design 18
Building Code & Fire Code Search

FIRE CODE REQUIREMENTS -- FIRE-RATED CONSTRUCTION continued

Checkmark each item to be done and cross out the check when completed. Mark with a -- if an item is not to be done. If an item is in doubt, mark with question mark and add a note of what to do to resolve the question. By: Dates:

__ FLOOR CONSTRUCTION
   Check: __ Broadscope drawings __ Details __ Specifications

__ FLOOR FINISHES
   Check: __ Broadscope drawings __ Details __ Specifications

__ ROOF CONSTRUCTION
   Check: __ Broadscope drawings __ Details __ Specifications
   Coord: __ Structural

__ ROOF FINISHES
   Check: __ Broadscope drawings __ Details __ Specifications

__ SKYLIGHTS AND MONITORS
   Check: __ Broadscope drawings __ Details __ Specifications

__ PARAPETS
   Check: __ Broadscope drawings __ Details __ Specifications

__ EXTERIOR STAIR CONSTRUCTION
   Check: __ Broadscope drawings __ Details __ Specifications
   Coord: __ Structural

__ INTERIOR SMOKE TOWER/FIRE STAIR CONSTRUCTION
   Check: __ Broadscope drawings __ Details __ Specifications
   Coord: __ Structural

__ ELEVATOR SHAFT WALLS
   Check: __ Broadscope drawings __ Details __ Specifications
   Coord: __ Structural

__ TRASH CHUTES
   Check: __ Broadscope drawings __ Details __ Specifications

__ ABOVE-CEILING FIRE BARRIERS
   Check: __ Broadscope drawings __ Details __ Specifications

__ ATTIC FIRE BARRIERS
   Check: __ Broadscope drawings __ Details __ Specifications

Add notes as needed after each checklisted task, such as:
Initials of who is to do it, when it's to start, when to review, who to coordinate with, and when it's to be finished.

Phase 3: Schematic Design 19
Building Code & Fire Code Search

FIRE CODE REQUIREMENTS -- EXITS, CORRIDORS, STAIRS, RAMPS, & COURTS

Checkmark each item to be done and cross out the check when completed. Mark with a -- if an item is not to be done. If an item is in doubt, mark with question mark and add a note of what to do to resolve the question.   By: Dates:

EXITS

___ Allowable sq. ft. per occupant
      Check: ___ Predesign ___ Schematics ___ Design Development

___ Broadcast Working Drawings
      Check: ___ Predesign ___ Schematics ___ Design Development

___ Occupants per floor
      Check: ___ Predesign ___ Schematics ___ Design Development

___ Exits required per floor
      Check: ___ Predesign ___ Schematics ___ Design Development

___ Distance between exits
      Check: ___ Predesign ___ Schematics ___ Design Development

EXIT CORRIDORS

___ Required minimum widths
      Check: ___ Design Development ___ Broadscope Working Drawings

___ Maximum dead end length
      Check: ___ Design Development ___ Broadscope Working Drawings

___ Maximum distance between exits
      Check: ___ Design Development ___ Broadscope Working Drawings

___ Required minimum ceiling height
      Check: ___ Design Development ___ Broadscope Working Drawings

___ Fire rating
      Check: ___ Broadscope drawings ___ Details ___ Specifications

Add notes as needed after each checklisted task, such as:
Initials of who is to do it, when it's to start, when to review, who to coordinate with, and when it's to be finished.

Phase 3: Schematic Design 20
Building Code & Fire Code Search

FIRE CODE REQUIREMENTS -- EXITS, CORRIDORS, STAIRS, RAMPS, & COURTS
continued

Checkmark each item to be done and cross out the check when completed. Mark with a -- if an item is not to be done. If an item is in doubt, mark with question mark and add a note of what to do to resolve the question. By: Dates:

EXIT STAIRS

___ Required minimum gross widths
   Check: ___ Design Development ___ Broadscope Working Drawings

___ Required minimum net widths
   Check: ___ Design Development ___ Broadscope Working Drawings

___ Required landings
   Check: ___ Design Development ___ Broadscope Working Drawings

___ Required minimum heights/clearances
   Check: ___ Design Development ___ Broadscope Working Drawings

___ Minimum and maximum risers
   Check: ___ Broadscope drawings ___ Details ___ Specifications

___ Minimum and maximum treads
   Check: ___ Broadscope drawings ___ Details ___ Specifications

___ Handrail heights
   Check: ___ Broadscope drawings ___ Details ___ Specifications

___ Corner panic barriers
   Check: ___ Broadscope drawings ___ Details ___ Specifications

___ Emergency battery powered lighting
   Check: ___ Broadscope drawings ___ Details ___ Specifications

___ Smoke ventilation
   Check: ___ Broadscope drawings ___ Details ___ Specifications

___ Acoustical treatment
   Check: ___ Broadscope drawings ___ Details ___ Specifications

___ Signage
   Check: ___ Broadscope drawings ___ Details ___ Specifications

___ Fire rating
   Check: ___ Broadscope drawings ___ Details ___ Specifications

Add notes as needed after each checklisted task, such as:
Initials of who is to do it, when it's to start, when to review, who to coordinate with, and when it's to be finished.

Phase 3: Schematic Design 21
Building Code & Fire Code Search

FIRE CODE REQUIREMENTS -- EXITS, CORRIDORS, STAIRS, RAMPS, & COURTS
continued

Checkmark each item to be done and cross out the check when completed. Mark with a -- if an item is not to be done.
If an item is in doubt, mark with question mark and add a note of what to do to resolve the question. By: Dates:

EXIT RAMPS

___ Required minimum gross widths
    Check: ___ Design Development ___ Broadscope Working Drawings

___ Required minimum net widths
    Check: ___ Design Development ___ Broadscope Working Drawings

___ Required landings
    Check: ___ Design Development ___ Broadscope Working Drawings

___ Maximum slope
    Check: ___ Design Development ___ Broadscope Working Drawings

___ Flooring
    Check: ___ Broadscope drawings ___ Details ___ Specifications

___ Required minimum heights/clearances
    Check: ___ Design Development ___ Broadscope Working Drawings

___ Handrail heights
    Check: ___ Broadscope drawings ___ Details ___ Specifications

___ Handicap railings
    Check: ___ Broadscope drawings ___ Details ___ Specifications

___ Emergency battery powered lighting
    Check: ___ Broadscope drawings ___ Details ___ Specifications

___ Signage
    Check: ___ Broadscope drawings ___ Details ___ Specifications

___ Fire rating
    Check: ___ Broadscope drawings ___ Details ___ Specifications

Add notes as needed after each checklisted task, such as:
Initials of who is to do it, when it's to start, when to review, who to coordinate with, and when it's to be finished.

Phase 3: Schematic Design 22
Building Code & Fire Code Search

FIRE CODE REQUIREMENTS -- EXITS, CORRIDORS, STAIRS, RAMPS, & COURTS continued

Checkmark each item to be done and cross out the check when completed. Mark with a -- if an item is not to be done. If an item is in doubt, mark with question mark and add a note of what to do to resolve the question. By: Dates:

EXIT COURTS AND PASSAGES

___ Required minimum gross widths
      Check: ___ Design Development ___ Broadscope Working Drawings

___ Required minimum net widths
      Check: ___ Design Development ___ Broadscope Working Drawings

___ Required minimum heights/clearances
      Check: ___ Design Development ___ Broadscope Working Drawings

___ Corner panic barriers
      Check: ___ Broadscope drawings ___ Details ___ Specifications

___ Emergency battery powered lighting
      Check: ___ Broadscope drawings ___ Details ___ Specifications

___ Smoke ventilation
      Check: ___ Broadscope drawings ___ Details ___ Specifications

___ Acoustical treatment
      Check: ___ Broadscope drawings ___ Details ___ Specifications

___ Signage
      Check: ___ Broadscope drawings ___ Details ___ Specifications

___ Fire rating
      Check: ___ Broadscope drawings ___ Details ___ Specifications

Add notes as needed after each checklisted task, such as:
Initials of who is to do it, when it's to start, when to review, who to coordinate with, and when it's to be finished.

Phase 4: Design Development 1

PROJECT MANAGEMENT CHECKLIST

Project Name/No: Notes by:

Dates Checked:

(The numbers beside the section titles below refer to the AIA SCOPE OF DESIGNATED SERVICES index numbers: AIA DOC. B162.)

ADMINISTRATION -- UPDATES AFTER THE SCHEMATIC PHASE

Checkmark each item to be done and cross out the check when completed. Mark with a -- if an item is not to be done. If an item is in doubt, mark with question mark and add a note of what to do to resolve the question. By: Dates:

___ Back-check and clear leftover tasks on the PHASE 3: SCHEMATIC DESIGN checklist.

___ Make a calendar schedule of future time, budget, and progress reviews.

___ Review scheduled dates for the final design presentation and later phases. Revise the schedule as needed.

___ Update the project planning chart.

___ Bring project records up to date by recording all pertinent discussions and decisions from the previous phase that haven't yet been recorded.

___ Update contact names, phone numbers, addresses, remarks, etc., in the Project Directory.

___ Input Project Directory updates into the office-wide Project Directory data base.

ADMINISTRATION -- PROJECT FINANCIAL MANAGEMENT

Checkmark each item to be done and cross out the check when completed. Mark with a -- if an item is not to be done. If an item is in doubt, mark with question mark and add a note of what to do to resolve the question. By: Dates:

___ Update the project office cost records.

___ Update the project's future work-hour and cost projections.

___ Review personnel allocation in light of latest money and time budgets.

___ Confirm the monthly statement submittal schedule and format with the client's bookkeeper.

___ Confirm the format and substantiating data required for submittal of monthly reimbursable statements. (Consider separating regular payment statements from reimbursable statements.)

___ Establish a schedule for documenting job costs in order to expedite submittals for payment to client.

Add notes as needed after each checklisted task, such as:
Initials of who is to do it, when it's to start, when to review, who to coordinate with, and when it's to be finished.

Phase 4: Design Development 2

DISCIPLINES COORDINATION AND DOCUMENT CHECKING

Checkmark each item to be done and cross out the check when completed. Mark with a -- if an item is not to be done. If an item is in doubt, mark with question mark and add a note of what to do to resolve the question. By: Dates:

See the CONSULTANT/ENGINEERING DRAWINGS CROSS-COORDINATION CHECKLIST section in PHASE 5: CONSTRUCTION DOCUMENTS for cross reference data.

___ Require all consultants to do their Design Development drawings according to the same scale, format, and drawing positioning as the architectural drawings. (Preferably, provide architectural base sheets of the overall building plans and sections and require consultants to do their work on transparency overlays.)

___ Identify any new consultants required for this phase, and negotiate contracts.

___ Before finalizing new consultant contracts, review service and contract terms with the client and obtain written client approvals.

___ Transmit updated information on building occupancies to consultants and make sure the architectural design team has the identical information.

___ Obtain or update the consultants' current estimates of building operating costs.

___ Review with the client the consultants' building operating cost estimates, and obtain written approval of the proposed mechanical and electrical systems.

___ Schedule group meetings to allow consultants to compare their drawings with one another. If interferences and contradictions can't be worked out on the spot, list the problems and schedule later meetings or calls to deal with them.

___ Review previous decisions on structural, construction, mechanical, and other building systems for possible economies and improvements. (Use formal Value Analysis techniques if you're trained in that process.)

___ Confirm that the various selected engineering and construction systems are compatible with one another.

___ Obtain updated estimates of spatial requirements for appurtenances and engineered systems.

AGENCY CONSULTING, REVIEW, AND APPROVALS

Checkmark each item to be done and cross out the check when completed. Mark with a -- if an item is not to be done. If an item is in doubt, mark with question mark and add a note of what to do to resolve the question. By: Dates:

___ Continue and update the data gathering in the checklists on regulatory agencies and codes. See the section on PERMITS AND APPROVALS in PHASE 2: SITE ANALYSIS, and the section on BUILDING CODE AND FIRE CODE SEARCH at the end of PHASE 3: SCHEMATIC DESIGN.

Add notes as needed after each checklisted task, such as:
Initials of who is to do it, when it's to start, when to review, who to coordinate with, and when it's to be finished.

Phase 4: Design Development 3

OWNER-SUPPLIED DATA COORDINATION

Checkmark each item to be done and cross out the check when completed. Mark with a -- if an item is not to be done. If an item is in doubt, mark with question mark and add a note of what to do to resolve the question. By: Dates:

___ Reconfirm the program's functional, occupancy and spatial requirements with the client.

___ Compare the developed design with the client's budget. Confirm the budget agreement or settle any contradictions between stated program needs and available funding.

___ Confirm the client's written approval of the schematic design and confirm that there is written agreement to proceed with Design Development.

___ Identify client preferences or requirements in types of construction bidding and contracting that might affect the format of construction drawings and specifications.

___ Obtain or update lists of special building equipment and fixtures required by the client that may affect consultants' work. Distribute the lists to the appropriate consultants.

ARCHITECTURAL DESIGN AND DOCUMENTATION

Checkmark each item to be done and cross out the check when completed. Mark with a -- if an item is not to be done. If an item is in doubt, mark with question mark and add a note of what to do to resolve the question. By: Dates:

___ Review any changes in the program and note their possible impact on the project design.

___ Review the schematic design, updates of the design, and changes in the program for possible violations of codes and regulations.

___ Review the schematic design, updates of the design, and changes in the program for possible conflicts with the original design intent or with fundamental engineering decisions.

___ If there are important differences between the present design and previous predesign, schematic, and engineering decisions, verify and document the reasons for and sources of the differences.

___ Submit a memo to all parties involved outlining the current status of work and the schedule for the Design Development phase.

___ Verify that all parties involved have received completely up-to-date program and schematic design data. Retrieve, or otherwise remove, leftover obsolete design and program information.

___ If there are changes in design staff between the Schematic phase and the Design Development phase, confirm that new staff members have acquired and assimilated all previous design data.

___ Confirm the type of construction contract to be used, such as single or separate contracts, and evaluate the effect of the contract type on drawing and specifications content and format.

Add notes as needed after each checklisted task, such as:
Initials of who is to do it, when it's to start, when to review, who to coordinate with, and when it's to be finished.

Phase 4: Design Development 4

ARCHITECTURAL DESIGN AND DOCUMENTATION   continued

Checkmark each item to be done and cross out the check when completed. Mark with a -- if an item is not to be done. If an item is in doubt, mark with question mark and add a note of what to do to resolve the question. By: Dates:

__ Prepare architectural Design Development drawings:

    __ Site Plan.
    __ Floor Plans.
    __ Roof Plan.
    __ Cross Sections.
    __ Exterior Elevations.
    __ Interior Elevations.
    __ Wall Sections.
    __ Design Details.

__ Prepare and coordinate Outline Specifications. See the section on SPECIFICATION WRITING AND COORDINATION at the end of PHASE FIVE: CONSTRUCTION DOCUMENTS for the specifications checklist.

__ Review architectural design development drawings as they are in process and compare them with the structural, mechanical, electrical, transportation, and other consultants' drawings by means of transparency overlays.

__ Schedule dates to confirm that the architectural designer or design team is fully informed of the most up to date consultants' information.

__ Schedule dates during Design Development, to compare the design as developing with budget, program, and regulatory requirements. Note any changes in building area, siting, structure, mechanical systems, construction systems, and materials that have occurred.

__ Determine and note reasons for changes in design. Review questionable changes with those who initiated them.

__ Prepare building floor area calculations, building volume, usable area ratios, and other numerical comparisons with program requirements.

__ Review preferred construction methods for impact on design and documentation.

__ Prepare data on costs and availability of special equipment and furnishings.

__ Confirm with the client whether a detailed construction cost estimate, such as a quantity survey, is desired with the design development package. (A detailed cost estimate, as opposed to the "Statement of Probable Construction Cost" is charged as an additional service, but is highly recommended.)

__ Submit Design Development documents (drawings, calculations, contracts, outline specifications, and updates on construction cost estimates) to the client. See PRESENTATIONS.

Add notes as needed after each checklisted task, such as:
Initials of who is to do it, when it's to start, when to review, who to coordinate with, and when it's to be finished.

Phase 4: Design Development 5

STRUCTURAL DESIGN AND DOCUMENTATION

Checkmark each item to be done and cross out the check when completed. Mark with a -- if an item is not to be done. If an item is in doubt, mark with question mark and add a note of what to do to resolve the question. By: Dates:

Also see the STRUCTURAL ENGINEERING checklist in the PREDESIGN AND PROGRAMMING section of PHASE 1: PREDESIGN.

___ Schedule structural and architectural coordination meetings.

___ Schedule structural, mechanical, and architectural drawing cross-checking meetings.

___ Review and reach agreement with the structural engineer on the number and content of structural Design Development documents.

   ___ Design criteria.
   ___ Structural grid or system.
   ___ Structural framing plan(s) and sections(s).
   ___ Preliminary foundation plan.
   ___ Estimated sizing of primary structural members.
   ___ Calculations.
   ___ Required clearances for other work.
   ___ Materials schedules.
   ___ Specifications.
   ___

___ Schedule completion dates for interim and final structural Design Development drawings and specifications.

___ Confirm with the structural engineer that the proposed structural system satisfies all legal requirements.

Add notes as needed after each checklisted task, such as:
Initials of who is to do it, when it's to start, when to review, who to coordinate with, and when it's to be finished.

Phase 4: Design Development 6

MECHANICAL DESIGN AND COORDINATION

Checkmark each item to be done and cross out the check when completed. Mark with a -- if an item is not to be done. If an item is in doubt, mark with question mark and add a note of what to do to resolve the question. By: Dates:

___ Schedule mechanical and architectural coordination meetings.

___ Schedule mechanical, structural, and architectural drawing cross-checking meetings.

___ Review and reach agreement with the mechanical engineer on the number and content of mechanical Design Development documents.

    ___ Design criteria:
        ___ Noise and vibration control.
        ___ HVAC system type and standard.
        ___ Fire protection systems.
        ___ Plumbing supply and drain types and standards.
    ———

    ___ Building plans and sections to show:
        ___ Equipment sizes and locations.
        ___ Chase sizes and locations.
        ___ Duct sizes and locations.
        ___ Mechanical equipment estimated spatial requirements in plan.
        ___ Mechanical equipment estimated spatial requirements in section.
        ___ HVAC calculations.
    ———

    ___ Energy use and conservation calculations.
    ___ Preliminary equipment and materials schedules.
    ___ Outline specifications.
    ———

___ Schedule completion dates for interim and final mechanical Design Development drawings and specifications.

___ Confirm with the mechanical consultant the acquisition of necessary approvals and permits for all utility services.

    ___ Gas.
    ___ Electric.
    ___ Water.
    ___ Sewer.
    ___ Communications.
    ———

___ Confirm with the mechanical consultant the compliance of the building mechanical and plumbing system design with codes and utility company requirements.

___ Acquire estimates for probable construction costs of the building's mechanical systems.

___ Acquire estimates for probable operating costs of the building's mechanical systems.

Add notes as needed after each checklisted task, such as:
Initials of who is to do it, when it's to start, when to review, who to coordinate with, and when it's to be finished.

Phase 4: Design Development 7

ELECTRICAL DESIGN AND DOCUMENTATION

Checkmark each item to be done and cross out the check when completed. Mark with a -- if an item is not to be done.
If an item is in doubt, mark with question mark and add a note of what to do to resolve the question. By: Dates:

___ Schedule electrical and architectural coordination meetings.

___ Schedule multidiscipline and architectural drawing cross-checking meetings.

___ Review and reach agreement with the electrical engineer on the number and content of electrical Design Development documents.

    ___ Building plans and sections to show:

        ___ Reflected ceiling lighting plans.
        ___ Power and switching.
        ___ Fire detection and alarm systems.
        ___ Security system.
        ___ Communications equipment, chases, and outlets.
        ___ Electrical equipment sizes, locations, and capacities.
        ___ Electrical vaults, transformer rooms.
        ___ Chase sizes and locations.
        ___ Duct sizes and locations.

___ Confirm and arrange the assistance of the electrical engineer in obtaining approvals and permits for electrical and communications services.

___ Schedule completion dates for interim and final electrical Design Development drawings and specifications.

___ Confirm with the engineer that the building electrical system design complies with codes and utility company requirements.

___ Obtain preliminary estimates for probable electrical systems construction costs.

Add notes as needed after each checklisted task, such as:
Initials of who is to do it, when it's to start, when to review, who to coordinate with, and when it's to be finished.

Phase 4: Design Development 8

CIVIL DESIGN AND DOCUMENTATION

Checkmark each item to be done and cross out the check when completed. Mark with a -- if an item is not to be done. If an item is in doubt, mark with question mark and add a note of what to do to resolve the question. By: Dates:

___ Confirm that results of all previously requested site tests have been received and transmitted to the client, consultants, and the design team.

___ Identify additional tests that may be required.

___ Update the Test Log and file.

___ Schedule civil, structural, and architectural coordination meetings.

CIVIL DESIGN AND DOCUMENTATION

Checkmark each item to be done and cross out the check when completed. Mark with a -- if an item is not to be done. If an item is in doubt, mark with question mark and add a note of what to do to resolve the question. By: Dates:

___ Schedule civil, structural, landscaping, and architectural drawing cross-checking meetings.

___ Review and reach agreement with the civil engineer on the number and content of civil Design Development documents.

   ___ Site plans and sections to show:

      ___ Cut and fill.
      ___ Excavations.
      ___ Irrigation.
      ___ Drainage.
      ___ Site related construction.
      ___

   ___ Outline specifications.

   ___

___ Schedule completion dates for interim and final civil Design Development drawings and specifications.

___ Check and confirm compliance of sitework and civil engineering design with codes and regulations.

___ Obtain preliminary estimates for probable civil engineering-related construction costs.

Add notes as needed after each checklisted task, such as:
Initials of who is to do it, when it's to start, when to review, who to coordinate with, and when it's to be finished.

Phase 4: Design Development 9

LANDSCAPE DESIGN AND DOCUMENTATION

Checkmark each item to be done and cross out the check when completed. Mark with a -- if an item is not to be done.
If an item is in doubt, mark with question mark and add a note of what to do to resolve the question. By: Dates:

___ Schedule landscaping, architectural coordination meetings.

___ Schedule multidiscipline drawing cross-checking procedures or meetings.

___ Review and reach agreement with the landscape architect on the number and content of landscape Design Development documents:

    ___ Preliminary landscaping planning.
    ___ Site-related plumbing work.
    ___ Site-related electrical work.
    ___

    ___ Outline specifications.
    ___

___ Schedule completion dates for interim and final landscape Design Development drawings and specifications.

___ Obtain preliminary estimates for probable landscaping development costs.

Add notes as needed after each checklisted task, such as:
Initials of who is to do it, when it's to start, when to review, who to coordinate with, and when it's to be finished.

Phase 4: Design Development 10

INTERIOR DESIGN AND DOCUMENTATION

Checkmark each item to be done and cross out the check when completed. Mark with a -- if an item is not to be done. If an item is in doubt, mark with question mark and add a note of what to do to resolve the question.   By: Dates:

___ Schedule interior design and architectural coordination meetings.

___ Review and reach agreement with the interior designer on the number and content of interior Design Development documents.

    ___ Preliminary interior partition landscaping.
    ___ Preliminary furniture planning.
    ___ Materials and finishes palette.
    ___ Color schedule.
    ___ Outline specifications.
    ___

___ Schedule completion dates for interim and final interior Design Development drawings and specifications.

___ Obtain preliminary estimates for probable costs of interior design furnishings and fixtures.

MATERIALS RESEARCH AND SPECIFICATIONS

Checkmark each item to be done and cross out the check when completed. Mark with a -- if an item is not to be done. If an item is in doubt, mark with question mark and add a note of what to do to resolve the question.   By: Dates:

For a more detailed checklist, see the section on SPECIFICATION WRITING AND COORDINATION in PHASE 5: CONSTRUCTION DOCUMENTS.

___ Start research on materials, equipment, fixtures, and building systems. Create a products and materials file or binder.

___ Start a list of primary first choices and alternative choices in materials, etc.

___ Start project outline specifications in coordination with schematic architectural drawings.

___ Review with consultants the extent of outline specifications required from them at this phase.

Add notes as needed after each checklisted task, such as:
Initials of who is to do it, when it's to start, when to review, who to coordinate with, and when it's to be finished.

Phase 4: Design Development 11

PROJECT DEVELOPMENT SCHEDULING

Checkmark each item to be done and cross out the check when completed. Mark with a -- if an item is not to be done.
If an item is in doubt, mark with question mark and add a note of what to do to resolve the question. By: Dates:

___ Create or update the job calendar of estimated phase starts and completions.

    ___ Design Development.
    ___ Construction Documents.
    ___ Bidding/Negotiation.
    ___ Contract Administration.
    ___ Post Construction
    ___

___ Distribute copies of the new or updated job calendar to all job participants.

___ Create a schedule for budget and progress reviews. (AIA Doc. F721, 723, F800 Series)

ESTIMATING PROBABLE CONSTRUCTION COST

Checkmark each item to be done and cross out the check when completed. Mark with a -- if an item is not to be done.
If an item is in doubt, mark with question mark and add a note of what to do to resolve the question. By: Dates:

For a detailed cost estimating checklist, see the section on CONSTRUCTION COST ESTIMATING in PHASE 3: SCHEMATIC DESIGN.

___ Have all consultants prepare construction cost estimates for their phases of work.

___ Prepare an overall construction cost estimate with a clearly stated contingency factor.

PRESENTATIONS

Checkmark each item to be done and cross out the check when completed. Mark with a -- if an item is not to be done.
If an item is in doubt, mark with question mark and add a note of what to do to resolve the question. By: Dates:

With participation or management by the Marketing Director.

___ List and schedule all Design Development presentations:

    ___ Interim presentations to client.
    ___ Financing agencies.
    ___ Regulatory agencies.
    ___ Advisory boards/committees.

___ Identify client preferences in design presentation media.

___ Review possible future client uses of design presentation material for project promotion. Design the presentation accordingly.

Add notes as needed after each checklisted task, such as:
Initials of who is to do it, when it's to start, when to review, who to coordinate with, and when it's to be finished.

Phase 4: Design Development 12

PRESENTATIONS continued

Checkmark each item to be done and cross out the check when completed. Mark with a -- if an item is not to be done. If an item is in doubt, mark with question mark and add a note of what to do to resolve the question. By: Dates:

___ Review possible office uses of presentation material for publicity, office brochure, presentation to other client prospects, etc.

___ Plan the scope and media for the design development presentation:

    ___ Sketch rendering.
    ___ Finish 2-D or perspective rendering.
    ___ Block models.
    ___ Detail models.
    ___ Computer printout and/or plots.
    ___ Slides.
    ___ Video.

___ Verify that presentation graphics will reproduce well in printed media.

___ Obtain client approval of the presentation media and approval of any added costs that might be involved.

___ Make a storyboard and mini-mock-up of the presentation. Send a memo on required graphic standards for the design presentation to design staff members.

___ Prepare presentation materials in such a form as to maximize possible reuse in working drawing production as well as later marketing. Keep solid line drawing on base sheets, for example, and do rendered entourage or design variations on overlays so as to keep original solid line work available for reuse.

___ Prepare finished presentation materials: drawings, models, slides, video, electronic media, etc.

___ Prepare for the client, presentations of optional design features and variations to compare and decide upon.

___ Finalize building floor area calculations, building volume, usable area ratios, and other numerical comparisons with program requirements.

___ Prepare presentation data on preferred construction methods.

___ Present the design development documents and cost data.

Add notes as needed after each checklisted task, such as:
Initials of who is to do it, when it's to start, when to review, who to coordinate with, and when it's to be finished.

Phase 4: Design Development 13

POST-PRESENTATION

Checkmark each item to be done and cross out the check when completed. Mark with a -- if an item is not to be done.
If an item is in doubt, mark with question mark and add a note of what to do to resolve the question. By: Dates:

___ Identify new changes in the design required by the client.

___ Note any extended repercussions from design changes, and review with the client any extensions of the Scope of Work or newly required changes in design service time and cost.

___ Identify any contradictions between requested design changes and the original design program or prior client/designer decisions. Review these with the client.

___ Obtain the client's written agreement to proceed with the next phase: Construction Documents.

___ Update the schedule and completion dates for working drawing and specification production.

___ Prepare and submit final billing for this phase of work as per the design service contract.

Add notes as needed after each checklisted task, such as:
Initials of who is to do it, when it's to start, when to review, who to coordinate with, and when it's to be finished.

Phase 5: Construction Documents 1
Working Drawings

PROJECT MANAGEMENT CHECKLIST

Project Name/No: Notes by:

Dates Checked:

ADMINISTRATION -- UPDATES AFTER DESIGN DEVELOPMENT

Checkmark each item to be done and cross out the check when completed. Mark with a -- if an item is not to be done. If an item is in doubt, mark with question mark and add a note of what to do to resolve the question.   By: Dates:

___ Back-check and clear left-over tasks from PHASE 4: DESIGN DEVELOPMENT.

___ Make a calendar schedule for future time, budget, and progress reviews.

___ Review previously scheduled dates for the working drawing phases. Revise the schedule as needed.

___ Update the project planning chart.

___ Bring project records up to date by recording all pertinent discussions and decisions from the previous phase that haven't yet been recorded.

___ Update contact names, phone numbers, addresses, remarks, etc. in the Project Directory.

___ Input Project Directory updates into the office-wide Project Directory data base.

ADMINISTRATION -- PROJECT FINANCIAL MANAGEMENT

Checkmark each item to be done and cross out the check when completed. Mark with a -- if an item is not to be done. If an item is in doubt, mark with question mark and add a note of what to do to resolve the question.   By: Dates:

___ Update the project office cost records.

___ Update the project's future work-hour and cost projections.

___ Review personnel allocation in light of latest money and time budgets.

___ Confirm the client's written approval of the final design development documents and that there is written agreement to proceed with the Construction Documents.

___ Confirm the monthly statement submittal schedule and format with the client's bookkeeper.

___ Confirm the format and substantiating data required for submittal of monthly reimbursable statements.

___ Establish a schedule for documenting job costs in order to expedite submittals for payment to client.

Add notes as needed after each checklisted task, such as:
Initials of who is to do it, when it's to start, when to review, who to coordinate with, and when it's to be finished.

Phase 5: Construction Documents 2
Working Drawings

ADMINISTRATION -- SCHEDULING AND PERSONNEL ALLOCATION

Checkmark each item to be done and cross out the check when completed. Mark with a -- if an item is not to be done.
If an item is in doubt, mark with question mark and add a note of what to do to resolve the question. By: Dates:

___ Confirm personnel previously scheduled for the working drawing phase. Schedule hiring as needed for future phases.

___ Estimate the final number of working drawing sheets. Calculate the average total allowable work hours per sheet based on the available fee.

___ Establish a clear chain of responsibility and command for the Construction Document phase. Confirm that no employee has more than one supervisor. Distribute a memo to all parties concerned.

___ Make a master list of personnel construction document assignments.

___ Schedule training sessions for personnel who are not experienced in special systematic production methods you use.

ADMINISTRATION -- WORKING DRAWING PLANNING

Checkmark each item to be done and cross out the check when completed. Mark with a -- if an item is not to be done.
If an item is in doubt, mark with question mark and add a note of what to do to resolve the question. By: Dates:

___ Verify that final design development and presentation drawings are available for reuse in the working drawing phase.

___ Review the working drawing sheet size, sheet module, and title block design. Confirm that the title block meets all special requirements of the client and of regulatory agencies.

___ Do an index of project drawings--architectural and consultant drawings. Decide the drawing numbering system:

   ___ Divisions by discipline.
   ___ CSI-related divisions.
   ___ Construction sequence divisions.
   ___

___ Do a one-fourth size sheet mini-mockup of all project drawings with sketches and/or notes of the data to go on each sheet. Distribute copies to concerned parties for review, then distribute final copies to all staff as a supervisory guide.

___ Decide the final printing system:

   ___ Print full-size or 1/2-size diazo, or both.
   ___ Print full-size or 1/2-size electrostatic copies.
   ___ Create originals at small scale as "full-size mini's."
   ___ Print on one or both sides of print sheets.
   ___ Print offset, black and white.
   ___ Print offset, multicolor.
   ___ Screen background information.
   ___

Add notes as needed after each checklisted task, such as:
Initials of who is to do it, when it's to start, when to review, who to coordinate with, and when it's to be finished.

Phase 5: Construction Documents 3
Working Drawings

ADMINISTRATION -- WORKING DRAWING PLANNING continued

Checkmark each item to be done and cross out the check when completed. Mark with a -- if an item is not to be done. If an item is in doubt, mark with question mark and add a note of what to do to resolve the question. By: Dates:

___ Decide specific drafting systems appropriate to parts or all of the project, and indicate them in the mini-mockup set.
    ___ Micro- or minicomputer text and/or CADD graphics.
    ___ Functional/simplified drafting.
    ___ Photodrafting.
    ___ Machine-made or paste-up titles.
    ___ Typed or computer printout notation.
    ___ Keynotes.
    ___ Standard notes.
    ___ Standard details.
    ___ Linked notes and/or details with CSI numbers.
    ___ Full sheets of reusable standard or typical file data.
    ___ Paste-up.
    ___ Freehand drafting.
    ___ Tape drafting.
    ___ Ink drafting.
    ___ Enlargement/reduction copying.
    ___ Tape drafting.
    ___ Stickybacks.
    ___ Base sheets and overlays.
    ___ Screened or solid line background sheets.

___ Identify the small-scope data that will be repeated in various drawings and should be copied in multiple for paste-up. Show the small-scope repetitive elements on the miniature working drawing mockup set.

___ Identify the large-scope data on working drawings that will be repeated in various sheets; identify them as "fixed" data suited to Base Sheets. Identify "variable" data, such as the different engineers' drawings that will be combined with architectural plans, and note them as Overlay Sheets.

___ Make a separate mockup of transparency sketches of base and overlay sheet combinations. Add base and overlay combinations to the original working drawing index, to show which base sheets and which overlays are combined to make complete final prints.

___ Review the completed working drawing index and the secondary index of base and overlay sheets with all concerned parties, for feedback and revisions.

___ Complete a matrix chart that lists the final planned working drawing sheets on one side and lists all base and overlay sheets on another. Make marks in the field of the matrix showing which base sheet and overlay(s) are combined to create each final working drawing sheet. Use codes or symbols to show which base sheets will be screened or reproduced in color in the final printing.

Add notes as needed after each checklisted task, such as:
Initials of who is to do it, when it's to start, when to review, who to coordinate with, and when it's to be finished.

Phase 5: Construction Documents 4
Working Drawings

ADMINISTRATION -- WORKING DRAWING PLANNING continued

Checkmark each item to be done and cross out the check when completed. Mark with a -- if an item is not to be done.
If an item is in doubt, mark with question mark and add a note of what to do to resolve the question. By: Dates:

___ Confirm the completion of limited architectural floor plan base sheet information for consultants' use. Such plan data may include:
    ___ Exterior walls and fenestration.
    ___ Interior walls and fixed partitions.
    ___ Door swings.
    ___ Equipment requiring plumbing hookups.
    ___ Fixtures and equipment requiring ventilation.
    ___ Major electrical spaces and equipment.
    ___ Reflected ceiling plans.
    ___ Mechanical and electrical chases.

DISCIPLINES COORDINATION AND DOCUMENT CHECKING

Checkmark each item to be done and cross out the check when completed. Mark with a -- if an item is not to be done.
If an item is in doubt, mark with question mark and add a note of what to do to resolve the question. By: Dates:

See the CONSULTANT/ENGINEERING DRAWINGS CROSS-COORDINATION CHECKLIST in this division for detailed cross reference checking data.

___ Require all consultants to do their working drawing plans and elevations according to the same scale, format, and drawing positioning as the architectural drawings. (Preferably, provide architectural base sheets of the overall building plans and sections and require consultants to do their work on transparency overlays.)

___ Identify any new consultants required for this phase, and negotiate contracts.

___ Before finalizing new consultant contracts, review service and contract terms with the client and obtain written client approval.

___ Transmit updated information on building occupancies to consultants; make sure the architectural design team has the identical updated information.

___ Obtain an update of the consultants' estimates of building operating costs.

___ Review with the client the consultants' building operating cost estimates; obtain from the client written approval of the proposed mechanical and electrical systems.

___ Schedule group meetings to allow consultants to compare their drawings with one another. If interferences and contradictions can't be worked out on the spot, list the problems and schedule later meetings or calls to deal with them.

Add notes as needed after each checklisted task, such as:
Initials of who is to do it, when it's to start, when to review, who to coordinate with, and when it's to be finished.

Phase 5: Construction Documents 5
Working Drawings

DISCIPLINES COORDINATION AND DOCUMENT CHECKING   continued

Checkmark each item to be done and cross out the check when completed. Mark with a -- if an item is not to be done. If an item is in doubt, mark with question mark and add a note of what to do to resolve the question. By: Dates:

__ Review previous decisions on structural, construction, mechanical, and other systems for possible economies and improvements.

__ Confirm that the various selected engineering and construction systems are compatible with one another.

__ Obtain updated estimates of spatial requirements for appurtenances and engineered systems.

__ Confirm that consultants, client, or others are handling the acquisition of approvals and permits for all utility services. (See section 5.03.)

    __ Gas.
    __ Electric.
    __ Water.
    __ Sewer.
    __ Telephone.
    __ Cable TV.
    __ Computer link.
    __ Utility-supplied steam or other heating medium.
    __ Utility-supplied cooling medium.
    __

__ Obtain or update lists of special building equipment and fixtures required by the client that may affect consultants' work. Distribute the lists to the appropriate consultants.

AGENCY CONSULTING, REVIEW, AND APPROVALS

Checkmark each item to be done and cross out the check when completed. Mark with a -- if an item is not to be done. If an item is in doubt, mark with question mark and add a note of what to do to resolve the question. By: Dates:

__ Establish a checklist and timetable for the client's applications for approvals and permits.

__ Continue and update the data gathering for the checklists on regulatory agencies and codes. See the section on PERMITS AND APPROVALS in PHASE 2: SITE ANALYSIS, and the section on BUILDING CODE AND FIRE CODE SEARCH at the end of PHASE 3: SCHEMATIC DESIGN.

Add notes as needed after each checklisted task, such as:
Initials of who is to do it, when it's to start, when to review, who to coordinate with, and when it's to be finished.

Phase 5: Construction Documents 6
Working Drawings

OWNER-SUPPLIED DATA COORDINATION

Checkmark each item to be done and cross out the check when completed. Mark with a -- if an item is not to be done.
If an item is in doubt, mark with question mark and add a note of what to do to resolve the question.   By: Dates:

___ Reconfirm the program's functional, occupancy, and spatial requirements with the client.

___ Compare the developed design with the client's budget. Confirm the budget agreement or settle any contradictions between stated program needs and available funding.

___ Confirm client preferences or requirements for types of construction bidding and contracting that might affect the format of construction drawings and specifications.

___ Identify possible or definite bid alternates and plan the content and organization of bid documents accordingly. See PHASE 6: PRE-BIDDING, BIDDING, AND NEGOTIATIONS.

ARCHITECTURAL DESIGN AND DOCUMENTATION

Checkmark each item to be done and cross out the check when completed. Mark with a -- if an item is not to be done.
If an item is in doubt, mark with question mark and add a note of what to do to resolve the question.   By: Dates:

Also see ADMINISTRATION -- WORKING DRAWING PLANNING.

___ Review any changes in the program, and note their possible impact on the project design.

___ Review the Design Development documents, updates of the design, and changes in the program for possible   violations of codes and regulations.

___ Review the Design Development documents, updates of the design, and changes in the program for possible   conflicts with the original design intent or with fundamental engineering decisions.

___ If there are significant differences between the present design and previous design and engineering decisions, verify and document the reasons for and sources of the differences.

___ Submit a memo to all involved parties outlining the current status of work and the schedule for the Construction Documents phase.

Add notes as needed after each checklisted task, such as:
Initials of who is to do it, when it's to start, when to review, who to coordinate with, and when it's to be finished.

Phase 5: Construction Documents 7
Working Drawings

ARCHITECTURAL DESIGN AND DOCUMENTATION continued

Checkmark each item to be done and cross out the check when completed. Mark with a -- if an item is not to be done. If an item is in doubt, mark with question mark and add a note of what to do to resolve the question.   By: Dates:

___ Verify that all involved parties have received completely up-to-date program and schematic design data. Retrieve or otherwise remove all hold-over, obsolete design and program information.

___ If there are changes in design staff between the Design Development phase and the Construction Document phase, confirm that new staff members have acquired and assimilated previous design data and understand the reasons for the present design solution.

___ Confirm the type of construction contract to be used, such as single or separate contracts, and evaluate the effect of the contract type on drawing and specifications content and format.

___ Prepare and coordinate final specifications. See the section on SPECIFICATION WRITING AND COORDINATION in this division of the manual for the specifications checklist.

___ Review architectural working drawings as they are in process and compare them with the structural, mechanical, electrical, transportation, and other consultants' drawings by means of transparency overlays.

___ Schedule coordination check points to confirm that the architectural production team is fully informed of the most up-to-date consultants' information.

___ Schedule dates to periodically compare the work as it has developed during the working drawing phase, with budget, program, and regulatory requirements. Note any changes in building area, siting, structure, mechanical systems, construction systems, and materials that have occurred.

___ Determine and note reasons for changes in the design. Review questionable changes with those who initiated them.

___ Review preferred construction methods for impact on design and documentation.

___ Prepare data on costs and availability of special equipment and furnishings.

___ Confirm with the client whether a detailed construction cost estimate, such as a quantity survey, is desired with the final working drawings. (A detailed cost estimate, as opposed to the "Statement of Probable Construction Cost," is charged as an additional service and is highly recommended.)

___ Confirm the date for submittal of all construction documents (drawings, calculations, contracts, specifications, and updates on construction cost estimates) to the client. See section on PRESENTATIONS.

Add notes as needed after each checklisted task, such as:
Initials of who is to do it, when it's to start, when to review, who to coordinate with, and when it's to be finished.

Phase 5: Construction Documents &
Working Drawings

STRUCTURAL DESIGN AND DOCUMENTATION

Checkmark each item to be done and cross out the check when completed. Mark with a -- if an item is not to be done.
If an item is in doubt, mark with question mark and add a note of what to do to resolve the question. By: Dates:

See the PRELIMINARY STRUCTURAL DECISIONS checklist in the PREDESIGN AND PROGRAMMING section of PHASE 1: PREDESIGN. For detailed coordination information, see the CROSS-COORDINATION CHECKLIST in this division.

___ Schedule phases of structural engineering document production and structural/architectural coordination meetings.

___ Schedule structural, mechanical, civil, and architectural drawing cross-checking meetings.

___ Review and reach agreement with the structural engineer on the number and content of structural Construction Documents.

    ___ Design criteria.
    ___ Structural grid or system.
    ___ Structural framing plan(s) and sections(s).
    ___ Foundation plan.
    ___ Calculations.
    ___ Required clearances for other work.
    ___ Structural details.
    ___ Materials schedules.
    ___ Specifications.
    ___

___ Schedule completion dates for interim and final structural working drawings and specifications.

___ Confirm with the structural engineer that the proposed structural system satisfies all legal requirements.

MECHANICAL DESIGN AND DOCUMENTATION

Checkmark each item to be done and cross out the check when completed. Mark with a -- if an item is not to be done.
If an item is in doubt, mark with question mark and add a note of what to do to resolve the question. By: Dates:

See also the CONSULTANT/ENGINEERING DRAWINGS CROSS-COORDINATION CHECKLIST in this division of the manual.

___ Establish mechanical documents production phases and dates for mechanical/architectural coordination meetings.

___ Schedule mechanical, structural, and architectural drawing cross-checking meetings.

Add notes as needed after each checklisted task, such as:
Initials of who is to do it, when it's to start, when to review, who to coordinate with, and when it's to be finished.

Phase 5: Construction Documents 9
Working Drawings

MECHANICAL DESIGN AND DOCUMENTATION continued

Checkmark each item to be done and cross out the check when completed. Mark with a -- if an item is not to be done.
If an item is in doubt, mark with question mark and add a note of what to do to resolve the question. By:	Dates:

___ Confirm with the mechanical consultant the acquisition of necessary approvals and permits for all utility services.

    ___ Gas.
    ___ Water.
    ___ Sewer.
    ___ Utility supplied steam or other heating medium.
    ___ Utility supplied cooling medium.
    ___

___ Review and reach agreement with the mechanical engineer on the number and content of final mechanical construction documents.

    ___ Building plans, sections, and other drawings to show:
        ___ Noise and vibration control.
        ___ HVAC system type(s) and standard(s).
        ___ Fire protection system(s).
        ___ Plumbing supply and drain types and standards.
        ___ Equipment sizes and locations.
        ___ Chase sizes and locations.
        ___ Duct sizes and locations.
        ___ Mechanical equipment spatial requirements in plan.
        ___ Mechanical equipment spatial requirements in section.
        ___ Mechanical fixture and equipment schedules.
        ___ Mechanical construction details.
    ___

    ___ HVAC heat load and cooling calculations.
    ___ Energy use and conservation calculations.
    ___ Equipment and materials schedules.
    ___ Specifications.
    ___ Mechanical systems operations and maintenance instructions.
    ___

___ Confirm with the mechanical consultant the compliance of the building mechanical and plumbing system design with codes and utility company requirements.

___ Identify changes in the scope of mechanical work that have occurred during the Design Development Phase.

___ Determine the impact on cost of revisions in mechanical work.

___ Confirm with the mechanical consultant the compliance of the building mechanical and plumbing system design with codes and utility company requirements.

___ Acquire estimates for probable construction costs of the building's mechanical systems.

___ Acquire estimates for probable operating costs of the building's mechanical systems.

Add notes as needed after each checklisted task, such as:
Initials of who is to do it, when it's to start, when to review, who to coordinate with, and when it's to be finished.

Phase 5: Construction Documents 10
Working Drawings

ELECTRICAL DESIGN AND DOCUMENTATION

Checkmark each item to be done and cross out the check when completed. Mark with a -- if an item is not to be done.
If an item is in doubt, mark with question mark and add a note of what to do to resolve the question. By: Dates:

See also the CONSULTANT/ENGINEERING DRAWINGS CROSS-COORDINATION CHECKLIST in this division of the manual.

___ Schedule electrical documents production phases and dates for electrical/architectural coordination meetings.

___ Schedule multidiscipline and architectural drawing cross-checking meetings.

___ Identify changes in the scope of electrical work that have occurred during the Design Development Phase.

___ Determine the impact on cost of revisions in electrical work.

___ Confirm that changes in the electrical design comply with legal requirements.

___ Review and reach agreement with the electrical engineer on the number and content of Electrical Construction Documents.

    ___ Building plans and sections to show:

        ___ Reflected ceiling lighting plans.
        ___ Power and switching.
        ___ Fire detection and alarm systems.
        ___ Security system.
        ___ Communications equipment, chases, and outlets.
        ___ Electrical equipment sizes, locations, and capacities.
        ___ Electrical vaults, transformer rooms.
        ___ Chase sizes and locations.
        ___ Duct sizes and locations.
        ___ Fixture schedules.
        ___ Electrical construction details.
        ___ Electrical, communications, security, fire, and related systems and equipment maintenance instructions.
        ___

    ___ Specifications.
    ___

___ Arrange the assistance of the electrical engineer in obtaining approvals and permits for electrical and communications services.

___ Obtain updated final estimates for probable electrical systems construction costs.

Add notes as needed after each checklisted task, such as:
Initials of who is to do it, when it's to start, when to review, who to coordinate with, and when it's to be finished.

Phase 5: Construction Documents 11
Working Drawings

CIVIL DESIGN AND DOCUMENTATION

Checkmark each item to be done and cross out the check when completed. Mark with a -- if an item is not to be done. If an item is in doubt, mark with question mark and add a note of what to do to resolve the question.   By: Dates:

___ Confirm that results of all previously requested site tests have been received and transmitted to the client, consultants, and the design team.

___ Identify additional tests that may be required.

___ Update the Test Log and file.

___ Schedule production phases and dates for civil/architectural coordination meetings.

___ Schedule civil, structural, landscaping, and architectural drawing cross-checking meetings.

___ Identify changes in the scope of civil engineering construction that have occurred through the Design Development Phase.

___ Determine the impact on cost of revisions in civil work.

___ Confirm that changes in the civil engineering design comply with legal requirements.

___ Review and reach agreement with the civil engineer on the number and content of civil engineering Construction Documents.

    ___ Site plans and sections to show:

        ___ Cut and fill.
        ___ Excavations.
        ___ Irrigation.
        ___ Drainage.
        ___ Site-related construction.
        ___ Civil engineering construction details.
        ———

    ___ Specifications.
    ———

___ Schedule completion dates for interim and final civil working drawings and specifications.

___ Check and confirm compliance of sitework and civil engineering design with codes and regulations.

___ Acquire updated estimates for probable civil engineering-related construction costs.

Add notes as needed after each checklisted task, such as:
Initials of who is to do it, when it's to start, when to review, who to coordinate with, and when it's to be finished.

Phase 5: Construction Documents 12
Working Drawings

LANDSCAPE DESIGN AND DOCUMENTATION

Checkmark each item to be done and cross out the check when completed. Mark with a -- if an item is not to be done.
If an item is in doubt, mark with question mark and add a note of what to do to resolve the question. By: Dates:

___ Schedule landscaping documents production phases and landscaping/architectural coordination meetings.

___ Schedule multidiscipline drawing cross-checking procedures or meetings.

___ Review and reach agreement with the landscape architect on the number and content of Landscape Construction Documents:

    ___ Landscaping plans.
    ___ Sitework construction details.
    ___ Site-related plumbing work.
    ___ Site-related electrical work.
    ___ Specifications.
    ___ Landscaping care instructions.

___ Identify special-order planting that must be ordered early, to assure delivery and installation before the completion date.

___ Schedule completion dates for interim and final landscape working drawings and specifications.

___ Update estimates for probable landscaping development costs.

INTERIOR DESIGN AND DOCUMENTATION

Checkmark each item to be done and cross out the check when completed. Mark with a -- if an item is not to be done.
If an item is in doubt, mark with question mark and add a note of what to do to resolve the question. By: Dates:

___ Establish production phases and schedule interior design/architectural coordination meetings.

___ List and schedule special-order furnishings (such as carpet) that must be ordered early, to assure delivery and installation before the move-in date.

___ Review and reach agreement with the interior designer on the number and content of interior Construction Documents:

    ___ Interior partition landscaping.
    ___ Furniture selection and planning.
    ___ Fixtures selection and finishes palette.
    ___ Materials and finishes palette.
    ___ Color schedule.
    ___ Interior design detailing.
    ___ Specifications.
    ___ Furnishings and finish material maintenance and cleaning instructions.

___ Schedule completion dates for the final interior drawings and specifications.

___ Update estimates for probable costs of interior design furnishings and fixtures.

Add notes as needed after each checklisted task, such as:
Initials of who is to do it, when it's to start, when to review, who to coordinate with, and when it's to be finished.

Phase 5: Construction Documents 13
Working Drawings

MATERIALS RESEARCH AND SPECIFICATIONS

For the detailed operations checklist, see the section on SPECIFICATION WRITING AND COORDINATION starting in this division of the manual.

PROJECT DEVELOPMENT SCHEDULING

Checkmark each item to be done and cross out the check when completed. Mark with a -- if an item is not to be done. If an item is in doubt, mark with question mark and add a note of what to do to resolve the question. By: Dates:

___ Create or update the job calendar of estimated phase starts and completions.

    ___ Construction Documents.

        ___ 10/20% review, independent quality control check.
        ___ Phase 1 @ _____ % completion.
        ___ Phase 2 @ _____ % completion.
        ___ 50% review, independent midpoint quality control check.
        ___ Phase 3 @ _____ % completion.
        ___ 80/90% review, independent quality control check.
    ___

        ___ Final checking and completion phase.

    ___ Bidding/Negotiation.
    ___ Contract Administration.
    ___ Post-construction.

___ Create a schedule for job budget and progress reviews. (AIA Doc. F721, 723, F800 Series)

___ Distribute copies of the new or updated job calendar to all job participants.

ESTIMATING PROBABLE CONSTRUCTION COST

Checkmark each item to be done and cross out the check when completed. Mark with a -- if an item is not to be done. If an item is in doubt, mark with question mark and add a note of what to do to resolve the question. By: Dates:

___ Obtain all consultants' final construction cost estimates.

___ Prepare an overall construction estimate of probable construction costs, with a clearly stated contingency factor.

Add notes as needed after each checklisted task, such as:
Initials of who is to do it, when it's to start, when to review, who to coordinate with, and when it's to be finished.

Phase 5: Construction Documents 14
Working Drawings

PRESENTATIONS

Checkmark each item to be done and cross out the check when completed. Mark with a -- if an item is not to be done.
If an item is in doubt, mark with question mark and add a note of what to do to resolve the question. By: Dates:

__ List and schedule all Construction Document presentations:

    __ Interim presentations to client.

    __ Presentations to financing agencies.

    __ Presentations to regulatory agencies.

    __ Presentations to advisory boards/committees.

    __

__ Review possible future client uses of working drawing material, such as base sheet floor plans for promotional graphics.

__ Review possible office uses of working drawing material for publicity, office brochure, presentation to other client prospects, etc.

__ Prepare presentation data on preferred construction methods.

__ Identify any last minute changes in the design required by the client.

__ Note any extended repercussions from design changes, and review with the client any extensions of the Scope of Work and any required changes in design service time and cost.

__ Identify any contradictions between requested design changes and the original design program or prior client/designer decisions. Review these with the client.

__ Obtain the client's written agreement to proceed with the next phase: PRE-BIDDING, BIDDING, AND NEGOTIATIONS.

__ Prepare and submit final billing for this phase of work as per the design service contract.

Add notes as needed after each checklisted task, such as:
Initials of who is to do it, when it's to start, when to review, who to coordinate with, and when it's to be finished.

Phase 5: Construction Documents 15
Consultant/Engineering Drawings Cross Coordination Checklist

PROJECT MANAGEMENT CHECKLIST

Basic architectural/consultant coordination and cross coordination are covered throughout the checklists in the SCHEMATIC DESIGN, DESIGN DEVELOPMENT, and CONSTRUCTION DOCUMENTS divisions of this manual.

These CONSULTANT/ENGINEERING DRAWINGS CROSS-COORDINATION CHECKLISTS provide detailed cross-coordination data and recommendations regarding which disciplines are responsible for which aspects of drawings and specifications. The coordination recommendations are compiled from procedural manuals of several well-experienced A/E firms, plus the California Council AIA's Production Office Procedures Manual.

Your specific requirements in terms of office policy or the needs of a particular project may differ from coordination recommendations in these checklists. Blank lines are provided for your modifications of the list.

MECHANICAL ENGINEERING CROSS COORDINATION -- PLUMBING

Project Name/No:

SPECIFICATIONS CONSTRUCTION COMPONENT:	DRAWINGS OR DWG. SCHEDULE BY:	BY:
___ Catch basins, meter traps, plumbing manholes, and related gratings, covers, and ladders.	___ MECHANICAL ___ ARCHITECTURAL ___ _____	___ MECHANICAL ___ _____
___ Underground tanks, concrete tunnels, and supports. Coordinated with:	___ MECHANICAL ___ _____ ___ STRUCTURAL	___ MECHANICAL ___ _____
___ Sleeves for plumbing pipe, other mechanical work.	___ MECHANICAL ___ ARCHITECTURAL ___ _____	___ MECHANICAL ___ _____
___ Ceiling and wall access doors/panels to plumbing not otherwise accessible.	___ MECHANICAL ___ ARCHITECTURAL ___ _____	___ MECHANICAL ___ _____

Add notes as needed after each checklisted task, such as:
Initials of who is to do it, when it's to start, when to review, who to coordinate with, and when it's to be finished.

Phase 5: Construction Documents 16
Consultant/Engineering Drawings Cross Coordination Checklist

MECHANICAL ENGINEERING CROSS COORDINATION -- PLUMBING continued

SPECIFICATIONS CONSTRUCTION COMPONENT:	DRAWINGS OR DWG. SCHEDULE BY:	BY:
___ Hangers, anchors, or other supports for plumbing related equipment.	___ MECHANICAL	___ MECHANICAL
Coordinated with:	___ STRUCTURAL	
___ Plumbing vents and flashing.	___ MECHANICAL ___ ARCHITECTURAL	___ MECHANICAL
___ Final painting of exposed plumbing.	N/A	___ ARCHITECTURAL
___ Wall- or floor-mounted bathroom/rest room accessories.	___ ARCHITECTURAL	___ ARCHITECTURAL
___ Floor plumbing trench.	___ ARCHITECTURAL	___ ARCHITECTURAL
Coordinated with:	___ STRUCTURAL	
___ Floor plumbing trench covers and gratings.	___ ARCHITECTURAL	___ ARCHITECTURAL
___ Floor drain.	___ MECHANICAL ___ ARCHITECTURAL	___ MECHANICAL
Coordinated with:	___ STRUCTURAL	

Add notes as needed after each checklisted task, such as:
Initials of who is to do it, when it's to start, when to review, who to coordinate with, and when it's to be finished.

Phase 5: Construction Documents 17
Consultant/Engineering Drawings Cross Coordination Checklist

MECHANICAL ENGINEERING CROSS COORDINATION -- HEATING, VENTILATING, AND AIR CONDITIONING

Project Name/No:

SPECIFICATIONS	CONSTRUCTION COMPONENT:	DRAWINGS OR DWG. SCHEDULE BY:	BY:
___	Floor mounts, bases, and pads for mechanical equipment.	___ ARCHITECTURAL ___ MECHANICAL	___ ARCHITECTURAL
	Coordinated with:	___ STRUCTURAL	
___	Floor trench.	___ ARCHITECTURAL ___ MECHANICAL	___ ARCHITECTURAL
	Coordinated with:	___ STRUCTURAL	
___	Floor trench covers and grating.	___ ARCHITECTURAL	___ ARCHITECTURAL
___	Wall and ceiling grilles, registers, and vents.	___ MECHANICAL	___ MECHANICAL
	Coordinated with:	___ ARCHITECTURAL	___ ARCHITECTURAL
___	Exterior wall louvers.	___ ARCHITECTURAL	___ ARCHITECTURAL
	Coordinated with:	___ STRUCTURAL ___ MECHANICAL	___ MECHANICAL
___	Door grilles.	___ ARCHITECTURAL	___ ARCHITECTURAL
	Coordinated with:	___ MECHANICAL	___ MECHANICAL
___	Sleeves for HVAC work.	___ MECHANICAL	___ MECHANICAL

Add notes as needed after each checklisted task, such as:
Initials of who is to do it, when it's to start, when to review, who to coordinate with, and when it's to be finished.

Phase 5: Construction Documents 18
Consultant/Engineering Drawings Cross Coordination Checklist

MECHANICAL ENGINEERING CROSS COORDINATION -- HEATING, VENTILATING, AND AIR CONDITIONING continued

SPECIFICATIONS CONSTRUCTION COMPONENT:	DRAWINGS OR DWG. SCHEDULE BY:	BY:
___ Ceiling and wall access doors/panels to HVAC work not otherwise accessible.	___ MECHANICAL ___ ARCHITECTURAL ___ _____	___ MECHANICAL ___ _____ ___ _____
___ Hangers, anchors, or other supports for HVAC-related equipment.	___ MECHANICAL ___ _____	___ MECHANICAL ___ _____
Coordinated with:	___ STRUCTURAL	
___ Gravity roof ventilators and power ventilators.	___ MECHANICAL ___ ARCHITECTURAL ___ _____	___ MECHANICAL ___ _____ ___ _____
Coordinated with:	___ STRUCTURAL	
___ Cooling tower supports.	___ STRUCTURAL ___ _____	___ STRUCTURAL ___ _____
___ Roof-mounted mechanical equipment supports.	___ STRUCTURAL ___ _____	___ STRUCTURAL ___ _____
___ Pitch pan or pocket flashing for roof-mounted mechanical equipment supports.	___ ARCHITECTURAL ___ _____	___ ARCHITECTURAL ___ _____
Coordinated with:	___ STRUCTURAL ___ _____	___ _____
___ Sheet metal and flashing for concrete bases at roof-supported mechanical equipment.	___ ARCHITECTURAL ___ _____	___ ARCHITECTURAL ___ _____
___ Curbs, nailers and flashing for roof-mounted fans, vents, and ductwork.	___ ARCHITECTURAL ___ _____	___ ARCHITECTURAL ___ _____

Add notes as needed after each checklisted task, such as:
Initials of who is to do it, when it's to start, when to review, who to coordinate with, and when it's to be finished.

Phase 5: Construction Documents 19
Consultant/Engineering Drawings Cross Coordination Checklist

MECHANICAL ENGINEERING CROSS COORDINATION -- HEATING, VENTILATING, AND AIR CONDITIONING continued

SPECIFICATIONS CONSTRUCTION COMPONENT:	DRAWINGS OR DWG. SCHEDULE BY:	BY:
___ Roof penetration counter-flashing for hot pipe, stacks, and ductwork.	___ MECHANICAL ___ ARCHITECTURAL	___ MECHANICAL
___ Final painting of mechanical equipment.	___ ARCHITECTURAL	

ELECTRICAL ENGINEERING CROSS COORDINATION

Project Name/No:

SPECIFICATIONS CONSTRUCTION COMPONENT:	DRAWINGS OR DWG. SCHEDULE BY:	BY:
___ Covers for underground electrical ducts and conduit.	___ ELECTRICAL	___ ELECTRICAL
___ Transformer vault.	___ STRUCTURAL ___ ARCHITECTURAL	___ STRUCTURAL
___ Electrical manholes, pull boxes, and ladders.	___ ELECTRICAL	___ ELECTRICAL
Coordinated with:	___ ARCHITECTURAL	
___ Electrical floor trench.	___ ARCHITECTURAL	___ ARCHITECTURAL
Coordinated with:	___ STRUCTURAL	
___ Electrical floor trench covers and gratings.	___ ARCHITECTURAL	___ ARCHITECTURAL
___ Electrical floor ducts.	___ ELECTRICAL	___ ELECTRICAL
Coordinated with:	___ STRUCTURAL	

Add notes as needed after each checklisted task, such as:
Initials of who is to do it, when it's to start, when to review, who to coordinate with, and when it's to be finished.

Phase 5: Construction Documents 20
Consultant/Engineering Drawings Cross Coordination Checklist

ELECTRICAL ENGINEERING CROSS COORDINATION continued

Project Name/No:

SPECIFICATIONS CONSTRUCTION COMPONENT:	DRAWINGS OR DWG. SCHEDULE BY:	BY:
___ Concrete base for switchboards, other electrical equipment.	___ ARCHITECTURAL	___ ARCHITECTURAL
Coordinated with:	___ STRUCTURAL	
___ Electrical conduit sleeves.	___ ELECTRICAL ___ ARCHITECTURAL	___ ELECTRICAL
Coordinated with:	___ STRUCTURAL	
___ Ceiling and wall access doors/panels to electrical equipment not otherwise accessible.	___ ELECTRICAL ___ ARCHITECTURAL	___ ELECTRICAL
___ Hangers, anchors, or other supports for electrical-related equipment.	___ ELECTRICAL	___ ELECTRICAL
Coordinated with:	___ STRUCTURAL	

STRUCTURAL ENGINEERING DRAWING AND ARCHITECTURAL COORDINATION

This coordination list is adapted from the Northern California Chapter, AIA Production Office Procedures Manual.

ARCHITECTURAL DRAWING	STRUCTURAL DRAWING
___ Structural grid with dimensions and coordinates.	___ Same as architectural.
___ Floor elevations at tops of slabs.	___ Same as architectural.
___ Foundation outline with overall building foundation dimensions.	___ Detailed foundation dimensions. Elevations of slabs, slab variations, and bottoms of footings.

Add notes as needed after each checklisted task, such as:
Initials of who is to do it, when it's to start, when to review, who to coordinate with, and when it's to be finished.

Phase 5: Construction Documents 21
Consultant/Engineering Drawings Cross Coordination Checklist

STRUCTURAL ENGINEERING DRAWING AND ARCHITECTURAL COORDINATION continued

ARCHITECTURAL DRAWING	STRUCTURAL DRAWING
___ Slab depressions, raised floors, and curbs, with locations and size dimensions. Details of mat frames, edge strips, etc.	___ Details of slabs and curbs showing location, and connection or integral design with other structural elements.
___ Openings in slabs and walls: location and size dimensions, details of frames, sills, jambs, edge strips, etc.	___ Details of structural elements related to openings such as integral beams, lintels, headers, etc.
___ Waterproofing membranes at slabs on or below grade, and below-grade walls.	___ Show relationship of membranes to structural members.
___ Outlines of structural elements such as columns, piers, pilasters, etc.	___ Identification and detail key or schedule key reference of structural elements. Locations and size and size dimensions of all structural elements.
___ Exterior construction: walks, ramps, slabs, steps, trenching, etc.--locations, dimensions, and architectural details.	___ Outline of architectural features and details of structural components. Details of connection of exterior construction to structural members. (Provide structural member details, sizes, and connection details to architect.)

Add notes as needed after each checklisted task, such as:
Initials of who is to do it, when it's to start, when to review, who to coordinate with, and when it's to be finished.

Phase 5: Construction Documents 22
Consultant/Engineering Drawings Cross Coordination Checklist

STRUCTURAL ENGINEERING DRAWING AND ARCHITECTURAL COORDINATION
continued

ARCHITECTURAL DRAWING	STRUCTURAL DRAWING
___ Steel frame fireproofing identification, notation, and detail key references.	___ Identification of primary, secondary, and non-support members to establish required fireproofing.
___ Miscellaneous metal details such as metal stairs, railings, ladders, etc.	___ Structural element connections to miscellaneous metal. (Provide structural member details, sizes, and connection details to architect.)
___ Details of anchors, hangers, brackets, etc.	___ Structural element connections, sleeves, and inserts for anchors, hangers, brackets etc. (Provide structural member details, sizes, and connection details to architect.)
___ Elevator shaft and machine beam support. (Location of beams supplied by elevator manufacturer.)	___ Locations and sizes of all elevator shaft and machine room supports.

Add notes as needed after each checklisted task, such as:
Initials of who is to do it, when it's to start, when to review, who to coordinate with, and when it's to be finished.

Phase 5: Construction Documents 23
Specification Writing and Coordination

PROJECT MANAGEMENT CHECKLIST

Project Name/No: Notes by:

Dates Checked:

ADMINISTRATION

Checkmark each item to be done and cross out the check when completed. Mark with a -- if an item is not to be done. If an item is in doubt, mark with question mark and add a note of what to do to resolve the question. By: Dates:

___ Confirm the type of construction contract.
      ___ Competitive Bidding -- Open
      ___ Competitive Bidding -- Selected Contractors.
      ___ Negotiated Contract.
      ___ Single Prime Contract.
      ___ Multiple Separate Contracts.
      ___ Stipulated Lump Sum. (AIA Doc. A101, A107)
      ___ Cost Plus Fee. (AIA Doc. A111, A117)

      Fee types:
      ___ Fixed Fee.
      ___ Fixed Fee with Guaranteed Maximum.
      ___ Percentage of Construction.
___

      Related options:
      ___ Phased Construction.
      ___ Fast Track.
      ___ Construction Management.
      ___ Design-Build.
      ___ Contractor prepared construction documents
___

___ Confirm the type of specification. (AIA HB 14)
      ___ Open/Contractor's Option.
      ___ Closed/Proprietary.
      ___ Product Approval.
      ___ Substitute Bid.
      ___ Approved Equal.
      ___ Product Description.
      ___ Performance.
      ___ Work Procedure.
___

___ Decide which specification section numbering system to use. The latest CSI five-digit number revisions are in the 1983 edition of Masterformat CSI MP-2-1-83. (Some master guide specifications are still numbered according to Masterformat CSI MP-2-1-78.)

Add notes as needed after each checklisted task, such as:
Initials of who is to do it, when it's to start, when to review, who to coordinate with, and when it's to be finished.

Phase 5: Construction Documents 24
Specification Writing and Coordination

ADMINISTRATION

Checkmark each item to be done and cross out the check when completed. Mark with a -- if an item is not to be done.
If an item is in doubt, mark with question mark and add a note of what to do to resolve the question. By: Dates:

___ Decide the specifications page numbering system. (Recommended: Number by division and division page number. Show start and finish page numbers for each division in the Table of Contents or Index.)

___ Establish type style, headings, and line indentation standards. (Recommended: Identify the project and design firm at the top of each page.)

___ Decide on any special reprographic combination of specifications with working drawing sheets:

    ___ Specifications printed on working drawing size sheets and included as the final sheets.
    ___ Specifications bound in with matching working drawing divisions.
    ___ Specifications bound with a detail book.
    ___

___ Review the design and production schedule to identify the best times for drawing/specification coordination checks and meetings.

___ Printing decisions:
    ___ Print on one or both sides of each sheet.
    ___ Paper -- type, quality, color code for major divisions.
    ___ Duplication method.
    ___ Binding -- fixed, looseleaf, looseleaf with sections fastened together.
    ___ Quantity.
    ___

PREDESIGN AND SCHEMATIC DESIGN PHASES

Checkmark each item to be done and cross out the check when completed. Mark with a -- if an item is not to be done.
If an item is in doubt, mark with question mark and add a note of what to do to resolve the question. By: Dates:

___ Schedule review meetings and/or drawing checks to coordinate decisions and alternatives on:
    ___ Room functions and relationships.
    ___ Construction system.
    ___ Structural system.
    ___ Mechanical system.
    ___ Lighting.
    ___ Dominant exterior materials.
    ___ Interior partitioning system.
    ___

    ___ Overall materials, finishes, and fixture quality.
        ___ Superior.
        ___ Middle Grade.
        ___ Economy Grade.
        ___ Mixed Grades.

Add notes as needed after each checklisted task, such as:
Initials of who is to do it, when it's to start, when to review, who to coordinate with, and when it's to be finished.

Phase 5: Construction Documents 25
Specification Writing and Coordination

DESIGN DEVELOPMENT PHASE

Checkmark each item to be done and cross out the check when completed. Mark with a -- if an item is not to be done. If an item is in doubt, mark with question mark and add a note of what to do to resolve the question. By: Dates:

___ Schedule review meetings and/or drawing checks to coordinate decisions on:
  ___ Preliminary room finish schedule.
  ___ Construction system.
  ___ Structural system.
  ___ Mechanical system.
  ___ Lighting.
  ___ Vertical transportation.
  ___ Exterior materials.
    ___ Site appurtenances.
    ___ Roofing.
    ___ Walls.
    ___ Fenestration.

  ___ Interior partition systems.
  ___ Cabinetry.
  ___ Specific area materials, finishes, and fixture quality.
    ___ Superior.
    ___ Middle Grade.
    ___ Economy Grade.
    ___ Mixed Grades.

RESEARCH AND DATA GATHERING

Checkmark each item to be done and cross out the check when completed. Mark with a -- if an item is not to be done. If an item is in doubt, mark with question mark and add a note of what to do to resolve the question. By: Dates:

___ Schedule review meetings and/or drawing checks to coordinate decisions on all construction materials and systems. (Recommended: Use the design development room finish schedule as a preliminary guide.)

___ Identify specification sections that can be completed early in the working drawing process.

___ Start a checklist of special standards and product literature required for this project.

___ Call or send for latest applicable product literature. See Sweet's Selection data file.

___ Send for the latest applicable testing agency standards.

___ Acquire copies of all applicable codes and regulations.

___ Acquire copies of previous relevant office specifications.

___ Contact the client for previous relevant specifications.

___ Create a Project Manual binder for preliminary organization of specification information. Use index tabs following the CSI Masterformat.

Add notes as needed after each checklisted task, such as:
Initials of who is to do it, when it's to start, when to review, who to coordinate with, and when it's to be finished.

Phase 5: Construction Documents 26
Specification Writing and Coordination

RESEARCH AND DATA GATHERING continued

Checkmark each item to be done and cross out the check when completed. Mark with a -- if an item is not to be done. If an item is in doubt, mark with question mark and add a note of what to do to resolve the question. By: Dates:

___ Create a master list of items to be specified. A list can be copied from the CSI Masterformat subdivisions. (If project management has used master planning and supervisory checklists such as those in the GUIDELINES PREDESIGN AND PLANNING MANUAL or THE GUIDELINES WORKING DRAWING PLANNING AND MANAGEMENT MANUAL, then all items to be detailed and specified are already identified along with their CSI coordination numbers.)

___ If using a master guide specification such as Masterspec2, Spectext, etc., match their completed sections to your master list. Note items to be covered that are not in the master guide specifications.

___ Identify specification sections that can be written or assembled first without extensive working drawing development.

___ Acquire printout or photocopy of overall master guide specification sections.

___ Cut and paste (physically or on computer) the preliminary reference or master guide specifications and add further notes as the guide for the typist or computer operator.

FINISH SPECIFICATIONS -- WRITING AND CHECKING

Checkmark each item to be done and cross out the check when completed. Mark with a -- if an item is not to be done. If an item is in doubt, mark with question mark and add a note of what to do to resolve the question. By: Dates:

___ Verify that specification section numbers consistently follow one system or one edition of a system such as the 1978 edition of the CSI Masterformat or the 1983 edition.

___ Assign proofreading and establish a proofreading system section by section. (Recommended: One person reads the last original draft aloud while another person reads and compares the verbal reading with the finished printout. This is the best way to catch errors and data drop outs.)

___ Check the completeness of specification sections. Check the consistency of the sequence of information in different sections. Items options, and sequence to check:
    ___ Materials.
        ___ Generic name.
        ___ Proprietary name with manufacturer.
        ___ Description by use.
        ___ Description by performance criteria.
        ___ Description by reference standard.

    ___ Required characteristics of materials.
        ___ Gauge or weight.
        ___ Sizes, nominal or finished.
        ___ Type of finish.
        ___ Allowable moisture content.

Add notes as needed after each checklisted task, such as:
Initials of who is to do it, when it's to start, when to review, who to coordinate with, and when it's to be finished.

Phase 5: Construction Documents 27
Specification Writing and Coordination

FINISH SPECIFICATIONS -- WRITING AND CHECKING continued

Checkmark each item to be done and cross out the check when completed. Mark with a -- if an item is not to be done. If an item is in doubt, mark with question mark and add a note of what to do to resolve the question. By: Dates:

___ Components or proportions of components of materials.
    ___ Mixes.
    ___ Temperature protection.
    ___ Moisture protection.

___ Installed location on the job if not fully indicated in the drawings.

___ Preparation for installation.
    ___ Pre-job inspection.
    ___ Coordination with other subcontractor(s).
    ___ Cleaning.
    ___ Preparation of surfaces.

___ Installation.
    ___ On-site fabrication.
    ___ Connection to other work.
    ___ Adjusting and fitting.
    ___ Finishing.

___ Coordination.
    ___ Broadscope working drawing sheet reference.
    ___ Detail drawing sheet reference.
    ___ Consultant's drawing sheet reference.
    ___ Related and/or connecting work by other trades or subcontractors.
    ___ Related other specifications sections.

___ Workmanship standards and tolerances.
    ___ Quantified measurements.
    ___ Referenced to published standards.
    ___ Approval by inspection.

___ Inspections and tests. (May be combined with workmanship standards and tolerances.)
___ Repair and patching.
___ Clean-up, preparation for other work.
___ Warranties, bonds, or guarantee requirements.
___ Postconstruction adjustments or service.

___ Read "Scope of Work" and "Work Not Included" articles in each section.

___ Verify all references to work in other sections.

___ Review the Special Conditions.

___ Distribute copies of specifications for content review by department heads and/or job captains, and the designated project site representative(s).

___ Check all final sets of printed specifications for incorrect collating, missing pages, and printing flaws, or blanks.

Add notes as needed after each checklisted task, such as:
Initials of who is to do it, when it's to start, when to review, who to coordinate with, and when it's to be finished.

Phase 5: Construction Documents 28
Specification Writing and Coordination

CHECKLIST -- COMMON OMISSIONS OR AMBIGUITIES IN SPECIFICATIONS

Checkmark each item to be done and cross out the check when completed. Mark with a -- if an item is not to be done. If an item is in doubt, mark with question mark and add a note of what to do to resolve the question. By: Dates:

The following items are described by contractors as the sources of the most common questions and trouble spots they find in specifications.

___ What is the precise procedure for getting approval of substitutions?

___ When separate contracts are involved, who pays for:
    ___ Heaters.
    ___ Emergency weather protection.
    ___ Storage sheds and platforms for subcontractors.
    ___ Cleanup and removal of subcontractors' trash.
    ___ Removal of faulty equipment.

___ Who sets survey marking lines, base lines, and elevations?

___ Who provides locations of utility hook up points?

___ Does a cash allowance cover purchase cost only, or does it include related costs of delivery, unloading, and storage?

___ When you say "work by others," which "others" do you mean?

___ When trades and subcontractors' work overlap, such as when tile work is specified as part of carpentry, or painting of equipment is part of HVAC, which subcontractors are responsible and which are not?

___ When equipment supports and anchors are specified for walls, floors, and ceilings, which subcontractor is responsible?

___ Who pays for tests and special inspections?

___ If tests are specified to be conducted in the presence of design firm representatives and they don't show up as scheduled, can that be taken as approval to proceed with testing without the observers?

___ When standards publications are identified as "the latest edition," does that mean the latest edition when the specifications were written, or the latest during bidding?

___ If a reference standard contradicts an article in the General Conditions, which should govern?

___ Is there a cut-off date when no further addenda will be sent out?

___ Is there a single list of all cash allowances that are scattered throughout the specifications?

___ Is there a single list of N.I.C. items, and items provided by Owner but installed by Contractor?

___ When specifying a new or unusual product or material, what is the name, address, and phone number of the source?

Add notes as needed after each checklisted task, such as:
Initials of who is to do it, when it's to start, when to review, who to coordinate with, and when it's to be finished.

Phase 6: Pre-bidding, Bidding, and Negotiations 1

PROJECT MANAGEMENT CHECKLIST

Project Name/No: Notes by:

Dates Checked:

ADMINISTRATION -- UPDATE ADMINISTRATION FROM CONSTRUCTION DOCUMENTS PHASE

Checkmark each item to be done and cross out the check when completed. Mark with a -- if an item is not to be done. If an item is in doubt, mark with question mark and add a note of what to do to resolve the question. By: Dates:

___ Back-check and clear up leftover tasks from the PHASE 5: CONSTRUCTION DOCUMENTS checklist.

___ Schedule any remaining final drawing checks.

___ Review previously scheduled dates for pre-bidding and bidding phase tasks. Revise the schedule as needed.

___ Update the project planning chart.

___ Bring project records up to date by recording all pertinent discussions and decisions from the previous phase that haven't yet been recorded.

___ Update contact names, phone numbers, addresses, remarks, etc., in the Project Directory.

___ Input Project Directory updates into the office-wide Project Directory data base.

ADMINISTRATION -- PRE-BID

Checkmark each item to be done and cross out the check when completed. Mark with a -- if an item is not to be done. If an item is in doubt, mark with question mark and add a note of what to do to resolve the question. By: Dates:

___ Obtain and confirm the client's written approval of the construction documents.

___ Obtain the client's written approval to proceed with the bidding process.

___ Establish whether the design firm or the client's legal representative is to identify special governing laws for out-of-state bidding, contracts, and construction.

___ Establish a tentative bid opening date and pre-bidding task time schedule.

___ Investigate whether other major projects are coming up for bid at the same time, and, if necessary, modify the bid date.

Add notes as needed after each checklisted task, such as:
Initials of who is to do it, when it's to start, when to review, who to coordinate with, and when it's to be finished.

Phase 6: Pre-bidding, Bidding, and Negotiations 2

ADMINISTRATION -- PRE-BID   continued

Checkmark each item to be done and cross out the check when completed. Mark with a -- if an item is not to be done.
If an item is in doubt, mark with question mark and add a note of what to do to resolve the question. By: Dates:

___ Prepare a plan and time schedule for assembling bid documents:
    ___ Bid Notice.
    ___ Bid Advertisement (if separate from the Invitation to Bid).
    ___ Invitation to Bid.
    ___ Instructions to Bidders. (AIA Doc. A701)
    ___ Contractor's Qualification Statement. (AIA Doc. A305)
    ___ Bid Form.
    ___ Owner-Contractor Agreement. (AIA Docs. A101, A107, A111)
    ___ Bid Documents Deposit.
    ___ Bid Security/Bid Bond. (AIA Doc. A310)
    ___ Performance Bond/Labor and Material Payment Bond. (AIA Doc. A311)
    ___ General and Supplementary Conditions. (AIA Doc. A201, A511)
    ___ Construction Documents: Drawings, Specifications, and Addenda.

CONSULTANT COORDINATION AND DOCUMENT CHECKING

Checkmark each item to be done and cross out the check when completed. Mark with a -- if an item is not to be done.
If an item is in doubt, mark with question mark and add a note of what to do to resolve the question. By: Dates:

___ Confirm the completion of document checking and coordination for the previous phase.

___ Identify alternates that concern the work of consultants.

___ Ask consultants to identify favored prospective subcontractors.

AGENCY CONSULTING, REVIEW, AND APPROVALS

Checkmark each item to be done and cross out the check when completed. Mark with a -- if an item is not to be done.
If an item is in doubt, mark with question mark and add a note of what to do to resolve the question. By: Dates:

___ Confirm that all necessary permits and approvals from regulatory agencies have been obtained.

___ Confirm that all necessary permits and approvals from public utilities have been obtained.

Add notes as needed after each checklisted task, such as:
Initials of who is to do it, when it's to start, when to review, who to coordinate with, and when it's to be finished.

Phase 6: Pre-bidding, Bidding, and Negotiations 3

OWNER-SUPPLIED DATA COORDINATION

Checkmark each item to be done and cross out the check when completed. Mark with a -- if an item is not to be done. If an item is in doubt, mark with question mark and add a note of what to do to resolve the question.   By: Dates:

___ Advise and confirm client's decision on the selection of contract type:

    ___ Competitive Bidding--Open.
    ___ Competitive Bidding--Selected Contractors.
    ___ Negotiated Contract.
    ___ Single Prime Contract.
    ___ Multiple Separate Contracts.
    ___ Stipulated Lump Sum. (AIA Doc. A101, A107)
    ___ Cost Plus Fee. (AIA Doc. A111, A117)

Fee types:

___ Fixed Fee.
___ Fixed Fee with Guaranteed Maximum.
___ Percentage of Construction.

Related options:

___ Phased Construction.
___ Fast Track.
___ Construction Management.
___ Design-Build.
___ Contractor-prepared construction documents.

___ If the selected contract form is Cost Plus Fee, establish the accounting and record keeping procedures to be used to monitor the contractor's performance.

___ If the project is out of state, consult with the client's legal counsel on the existence of any special laws regarding the bidding process, construction documents, and forms of agreement.

___ Identify and confirm the design firm and client's separate responsibilities in advertising for bids, receiving bids, negotiation, and acceptance.

___ Identify insurance coverage the client should have prior to the execution of the contract. (AIA Doc. A201)

___ Identify insurance and bonds the client will require the contractor to have. (AIA Doc. G610)

___ List and confirm what materials, equipment, and furnishings are supplied by the client and installed by the contractor.

___ List and confirm what materials, equipment, and furnishings are supplied by the client and installed by anyone other than the contractor.

___ Confirm that the client has supplied an accurate site survey, site legal description, and a soil and subsurface condition report, all to be included with the construction documents.

Add notes as needed after each checklisted task, such as:
Initials of who is to do it, when it's to start, when to review, who to coordinate with, and when it's to be finished.

Phase 6: Pre-bidding, Bidding, and Negotiations 4

BIDDING MATERIALS -- FOR NON-OPEN BIDDING, NEGOTIATED CONTRACT

Checkmark each item to be done and cross out the check when completed. Mark with a -- if an item is not to be done.
If an item is in doubt, mark with question mark and add a note of what to do to resolve the question. By: Dates:

___ Establish the criteria for contractor qualifications and acceptable contract terms.

___ Review selection(s) of preferred contractor(s) and discuss choices with the client.

___ Establish terms for bargaining and acceptable alternatives in scheduling and budgeting construction.

BIDDING MATERIALS -- FOR INVITED BIDDING

Checkmark each item to be done and cross out the check when completed. Mark with a -- if an item is not to be done.
If an item is in doubt, mark with question mark and add a note of what to do to resolve the question. By: Dates:

___ Establish the criteria for qualifications of acceptable contractors.

___ Select the preferred contractor(s) for negotiation, and review the selection with the client for approval.

___ Create invited bidders list.

___ Notify invited bidders.

BIDDING MATERIALS -- FOR OPEN BIDDING

Checkmark each item to be done and cross out the check when completed. Mark with a -- if an item is not to be done.
If an item is in doubt, mark with question mark and add a note of what to do to resolve the question. By: Dates:

___ Finish Instructions to Bidders and/or advertising for bidders.

___ Review bidding opening date and location with the office principal in charge.

___ Verify that the correct bid opening date, time, and place are included in the Invitation to Bidders and bid advertising.

___ Begin and maintain a Register of Bid Documents. (AIA Doc. G804)

___ Identify favored prospective prime contract bidders.

___ Identify media for bid advertising: construction periodical(s), newspapers, plan rooms, etc.

___ Write the criteria for acceptable bidder qualifications, and confirm criteria with the client. (AIA Doc. A305)

___ Write the Invitation to Bid, and obtain client's approval.

___ Identify the surety or bid bond to be required of bidders. (AIA Doc. A310, A311)

Add notes as needed after each checklisted task, such as:
Initials of who is to do it, when it's to start, when to review, who to coordinate with, and when it's to be finished.

Phase 6: Pre-bidding, Bidding, and Negotiations 5

BIDDING MATERIALS -- FOR OPEN BIDDING   continued

Checkmark each item to be done and cross out the check when completed. Mark with a -- if an item is not to be done. If an item is in doubt, mark with question mark and add a note of what to do to resolve the question.   By: Dates:

___ Publish and distribute the Invitation to Bid.

___ Obtain statements of qualification from prospective bidders.

___ Inform the client on the degree of response to bid advertising.

___ Notify selected bidders.

___ Identify and list all bid documents to be distributed to bidders, and identify the amount of bid document deposit.

___ Review the complete bid package with the client.

___ Identify the quantity of bid documents to distribute to each bidder and the total number of bidders.

___ Identify those other than bidders who will receive bid documents, such as the client, consultants, design firm staff members, and management.

___ Establish a document printing and distribution schedule.

___ Distribute bid documents to bidders, plan rooms, client, and all other concerned parties.

___ Schedule a pre-bid conference to review documents with prospective bidders.

___ Maintain a log of distributed documents, including bidders' deposit/security payments and refunds.

___ Route deposit/security payments to the accounting department.

___ Record refunds of deposits/security payments to disqualified bidders or bidders who withdraw.

___ Review proposed substitutions according to formal procedures established in the Instructions to Bidders.

___

Add notes as needed after each checklisted task, such as:
Initials of who is to do it, when it's to start, when to review, who to coordinate with, and when it's to be finished.

Phase 6: Pre-bidding, Bidding, and Negotiations 6

ADDENDA

Checkmark each item to be done and cross out the check when completed. Mark with a -- if an item is not to be done. If an item is in doubt, mark with question mark and add a note of what to do to resolve the question. By: Dates:

___ Prepare an addendum log in the Register of Bid Documents.

___ Distribute addenda to all bidders according to procedures established in the Bid Documents.

___ When responding to any bidder's request for clarification or additional data, send copies of the clarification data as an addendum to all other bidders.

BIDDING AND NEGOTIATIONS

Checkmark each item to be done and cross out the check when completed. Mark with a -- if an item is not to be done. If an item is in doubt, mark with question mark and add a note of what to do to resolve the question. By: Dates:

___ Hold pre-bid meeting(s) with prospective bidders and client.

___ Prepare a report on the pre-bid meeting(s) and send copies to all concerned parties.

___ Prepare a Bid Tabulation Form.

___ Remind prospective bidders of the bid opening data, time, and location, and confirm their participation.

___ Receive bids according to procedures in the Instructions to Bidders.

___ Check all bids to confirm the validity of the contractors' and major subcontractors' licenses.

___ Confirm that the rules regarding bid security are enforced.

ANALYSIS OF ALTERNATES AND SUBSTITUTIONS

Checkmark each item to be done and cross out the check when completed. Mark with a -- if an item is not to be done. If an item is in doubt, mark with question mark and add a note of what to do to resolve the question. By: Dates:

___ Create a "Confirmation Form" memo to set down in writing all verbal interpretations, instructions, and confirmations. Establish a time limit in which copies of such memos must be distributed.

___ Establish a record of consultations with the client on changes and alternates with space for notes confirming client approvals of alternates.

___ Establish a record of notifications to the Contractor(s) of approved and not approved alternates.

___ Notify all bidders of accepted substitutions.

Add notes as needed after each checklisted task, such as:
Initials of who is to do it, when it's to start, when to review, who to coordinate with, and when it's to be finished.

Phase 6: Pre-bidding, Bidding, and Negotiations 7

SPECIAL BIDDING SERVICES

Checkmark each item to be done and cross out the check when completed. Mark with a -- if an item is not to be done. If an item is in doubt, mark with question mark and add a note of what to do to resolve the question. By: Dates:

___ Identify special services from previous Scope of Work agreement.

___ List any added special bidding services desired by the client.

___ List and schedule any special negotiations that may be required.

___ Assist the client in establishing criteria and schedules for phased construction or multiple contracts.

BID EVALUATION

Checkmark each item to be done and cross out the check when completed. Mark with a -- if an item is not to be done. If an item is in doubt, mark with question mark and add a note of what to do to resolve the question. By: Dates:

___ Analyze the bids; check for errors or omissions.

___ Write a comparison of bid tabulations with the latest design firm construction cost estimates.

___ Review significant discrepancies between the bid tabulations and the last previous construction cost estimate.

___ Prepare a memo or report on the reasons for bid and estimate discrepancies, their impact, and recommended next steps.

___ Review cost and bid problems with project principals and the client.

___ Notify the client of bid expiration dates.

___ Review bids with the client and advise on bid acceptance and rejection. Obtain the client's acceptance of a bid or rejection of all bids.

___ If all bids are rejected, confer with the client to establish the next step of bidding invitations or negotiations.

___ Record reasons for acceptance/rejections in the project diary.

___ Advise the client how to draft a notice of acceptance that states an intent to execute the contract without specifically awarding the contract.

___ Notify all bidders of acceptance or rejection. If bidders are not present at bid opening, send them a tabulation of all bids within ten days of bid opening.

___ Receive returned documents and refund the bid deposits/security to unsuccessful bidders.

___ If the client wants to proceed with limited interim construction prior to awarding the final contract, advise the client on the form and content of this type of letter of intent. (Avoid participating in actual preparation of such letters -- that must be restricted to the client's legal counsel.)

Add notes as needed after each checklisted task, such as:
Initials of who is to do it, when it's to start, when to review, who to coordinate with, and when it's to be finished.

Phase 6: Pre-bidding, Bidding, and Negotiations 8

CONSTRUCTION CONTRACT AGREEMENTS

Checkmark each item to be done and cross out the check when completed. Mark with a -- if an item is not to be done. If an item is in doubt, mark with question mark and add a note of what to do to resolve the question. By: Dates:

__ Advise the client on construction contract format and content.

__ Have consultants assist on preparation of separate prime contracts.

__ Provide the client with a checklist of separate designer/client/contractor responsibilities as stated in the contract.

__ Advise client and contractor of their insurance responsibilities. Schedule times for confirmation of required insurance coverage.

__ Obtain performance and labor and material payment bonds from the contractor. Review and forward copies of bonds to client.

__ Obtain the contractor's certificate of insurance. Review and forward copies of certificate to client.

__ Obtain the client's property insurance policy. Review and forward copies to the contractor.

__ Identify and review the establishment of any non-typical insurance arrangements between the client and the contractor. Include descriptions of such arrangements in the contract.

__ Obtain the post-bid information from the accepted contractor as required in the Instructions to Bidders. (AIA Doc. A701)

__ Review the construction plan and time schedule with the client and contractor for inclusion in the contract.

__ Consult and assist with the client in negotiating and executing the final contract.

__ Prepare and send to the contractor(s) notices to proceed with the work.

POST-BIDDING ADMINISTRATION

Checkmark each item to be done and cross out the check when completed. Mark with a -- if an item is not to be done. If an item is in doubt, mark with question mark and add a note of what to do to resolve the question. By: Dates:

__ Create a log for recording all change orders and modifications to the contract. (AIA Doc. A G701)

__ Provide all necessary contract documents, specified equipment brochures, and related project data to the contractor.

__ Identify bid tabulation data, special agreements addenda, and memos, reports, minutes, and correspondence that should be included in the final Project Manual as part of construction contracts or construction documents.

__ Obtain written approval from the client to proceed with construction and construction administration.

Add notes as needed after each checklisted task, such as:
Initials of who is to do it, when it's to start, when to review, who to coordinate with, and when it's to be finished.

Phase 7: Construction Contract Administration 1

PROJECT MANAGEMENT CHECKLIST

Project Name/No: Notes by:

Dates Checked:

PROJECT ADMINISTRATION -- UPDATES AFTER BIDDING

Checkmark each item to be done and cross out the check when completed. Mark with a -- if an item is not to be done.
If an item is in doubt, mark with question mark and add a note of what to do to resolve the question. By: Dates:

___ Update the routing list of all parties who should receive memos and notices regarding project modifications, special instructions to the contractor, interpretations, clarifications, etc.

___ Back-check and clear up leftover tasks from the PHASE 6: PRE-BIDDING, BIDDING, AND NEGOTIATIONS checklist.

___ Make a calendar schedule of future time, budget, and progress reviews.

___ Review previously scheduled dates for the construction phases. Revise the schedule as needed.

___ Update the project planning chart.

___ Bring project records up to date by recording all pertinent discussions and decisions from the previous phase that haven't yet been recorded.

___ Update contact names, phone numbers, addresses, remarks, etc., in the Project Directory.

___ Input Project Directory updates into the office-wide Project Directory data base.

PROJECT ADMINISTRATION -- PRECONSTRUCTION

Checkmark each item to be done and cross out the check when completed. Mark with a -- if an item is not to be done.
If an item is in doubt, mark with question mark and add a note of what to do to resolve the question. By: Dates:

___ Create a Construction Contract Administration Manual with:

    ___ Overall Task Modules. (GUIDELINES)
    ___ Schedules and Progress Charts. (AIA Doc. A201, pgr. 4.11)
    ___ Construction Administration Checklist. (GUIDELINES)

    ___ Project Field Observations and Field Reports. (AIA Doc. G711)
    ___ Project Photo Surveys.

    ___ Client Approvals. (AIA Doc. A201)
    ___ Waivers, Receipts, and Vouchers. (AIA Doc. G706, G706A)

    ___ Field Orders.
    ___ Change Orders. (AIA Doc. G701, C701/CM, G709)
    ___ Supplemental Documents and Instructions. (AIA Doc. G710)

Add notes as needed after each checklisted task, such as:
Initials of who is to do it, when it's to start, when to review, who to coordinate with, and when it's to be finished.

Phase 7: Construction Contract Administration 2

PROJECT ADMINISTRATION -- PRECONSTRUCTION continued

Checkmark each item to be done and cross out the check when completed. Mark with a -- if an item is not to be done. If an item is in doubt, mark with question mark and add a note of what to do to resolve the question. By: Dates:

    ___ Tests. (AIA Doc. A201, pgr. 7.8)
    ___ Shop Drawings and Samples. (AIA Doc. G712)
    ___ Inspections, Permits and Approvals. (AIA Doc. A201, pgr. 7.8)

    ___ Certificates of Payment. (AIA Doc. G702)
    ___ Owner-Architect Agreement/Owner-Contractor Agreement.
    ___ Schedule of Values. (AIA HBC 13, G702, G702A)

    ___ Observations of Contractor Performance.
    ___ Certificate of Insurance. (AIA Doc. G705)
    ___ Record of Document Distribution. (AIA Doc. G810)

    ___ Construction Detail Jobsite Feedback.
    ___ Broadscope Working Drawing Jobsite Feedback.
    ___ Specifications Jobsite Feedback.

    ___ Final Inspections and Close-out.

___ Confirm the method and degree of contract administration and site observation.

    ___ Administration by Principal in Charge, Project Manager, Project Architect, other management or staff.
    ___ Part-time site observation visits. Estimate frequency and duration.
    ___ Full-Time Project Representative (AIA Doc. B352).

___ Assign and schedule construction administration personnel. Establish their hierarchy of command, communication, and responsibilities.

___ Design and schedule fabrication and installation of the project identification sign.

___ Design the construction barrier fence. Obtain client approval of the barrier design and colors.

PROJECT ADMINISTRATION -- ACTIONS REGARDING CONTRACTOR

Checkmark each item to be done and cross out the check when completed. Mark with a -- if an item is not to be done. If an item is in doubt, mark with question mark and add a note of what to do to resolve the question. By: Dates:

___ Confirm that the construction contract is complete, including the Schedule of Values.

___ Confirm that the contractor's Performance Bond and the Labor and Material Bond are correct. Make file copies and forward the originals to the client. (AIA Doc. A311)

___ Send the contractor a copy of the client's directions about required insurance.

___ Confirm that the contractor has filed the Certificate of Insurance with the client. (AIA Doc. G705)

___ Confirm that the contractor has acquired and paid for all necessary permits.

Add notes as needed after each checklisted task, such as:
Initials of who is to do it, when it's to start, when to review, who to coordinate with, and when it's to be finished.

Phase 7: Construction Contract Administration 3

PROJECT ADMINISTRATION -- ACTIONS REGARDING CONTRACTOR

Checkmark each item to be done and cross out the check when completed. Mark with a -- if an item is not to be done. If an item is in doubt, mark with question mark and add a note of what to do to resolve the question. By: Dates:

___ Acquire the list of proposed subcontractors from the prime contractor.

___ Ask consultants for their opinions regarding proposed subcontractors.

___ Assist the client in approving, disapproving, or acquiring more information on subcontractors proposed by the prime contractor. Obtain client approvals in writing. (AIA Doc. G805)

___ Send written rejection memos regarding disapproved subcontractors to the prime contractor.

___ Acquire the names of substitute subcontractors plus differences in construction time or cost caused by any substitution.

___ Assist the client in evaluating substitute contractors. Obtain client approvals in writing.

___ Write Change Orders when necessary to modify contract terms because of any substitution of subcontractors.

___ Review and approve, or have corrected, the contractor's Schedule of Values before the first scheduled application for payment.

___ Review and modify as necessary the contractor's schedule of required shop drawings, samples, and colors.

___ Review and approve, or have corrected, the contractor's estimated job construction progress schedule.

___ Notify the client of the estimated job construction schedule and review any scheduling problems.

___ Establish a tentative job observation schedule based on the contractor's estimated construction schedule.

___ Select the format for the Project Schedule chart:

  ___ Bar Chart.
  ___ CPM.
  ___ PERT.
  ___ Precedence Network.
  ___

___ Confirm that all contractor schedules (shop drawing, values, job progress, etc.) conform to contract requirements.

___ Distribute copies of job progress and observation schedules to the client, consultants, appropriate in-house staff members and management.

___ Provide the contractor with all necessary construction documents.

___ Establish a time for the preconstruction meeting.

Add notes as needed after each checklisted task, such as:
Initials of who is to do it, when it's to start, when to review, who to coordinate with, and when it's to be finished.

Phase 7: Construction Contract Administration 4

CONSULTANTS COORDINATION AND DOCUMENT CHECKING

Checkmark each item to be done and cross out the check when completed. Mark with a -- if an item is not to be done.
If an item is in doubt, mark with question mark and add a note of what to do to resolve the question. By: Dates:

___ Notify the consultants of selected prime contractor(s) and subcontractors.

___ Establish a time and coordinate the preconstruction meeting with the appropriate consultant representatives.

AGENCY CONSULTING, REVIEW, AND APPROVALS

Checkmark each item to be done and cross out the check when completed. Mark with a -- if an item is not to be done.
If an item is in doubt, mark with question mark and add a note of what to do to resolve the question. By: Dates:

___ Establish inspection schedules and records and coordinate with agency inspection requirements checklisted under CONSTRUCTION INSPECTION, SUBSTANTIAL COMPLETION AND PROJECT CLOSE-OUT.

OWNER-SUPPLIED DATA COORDINATION

Checkmark each item to be done and cross out the check when completed. Mark with a -- if an item is not to be done.
If an item is in doubt, mark with question mark and add a note of what to do to resolve the question. By: Dates:

___ Confirm that the client has applied for all necessary permanent utility services.

___ Confirm that the client has all necessary property insurance and has provided copies to the contractor.

___ Confirm with the client any previous understandings about contingency budgets (from 2% to 5% of construction cost) to allow for unavoidable changes or delays in the work.

___ Establish whether the client or the contractor will be responsible for purchasing property insurance. Confirm that all parties have reached written agreement as to how property insurance is to be handled. (AIA Doc. G701)

___ Identify any special hazards insurance to be included in the property insurance, and confirm written agreement between the client and contractor.

___ Schedule and coordinate the preconstruction meeting with the client.

___ Schedule and coordinate client-supplied materials, equipment, and furnishings.

Add notes as needed after each checklisted task, such as:
Initials of who is to do it, when it's to start, when to review, who to coordinate with, and when it's to be finished.

Phase 7: Construction Contract Administration 5

OFFICE CONSTRUCTION ADMINISTRATION

Checkmark each item to be done and cross out the check when completed. Mark with a -- if an item is not to be done. If an item is in doubt, mark with question mark and add a note of what to do to resolve the question. By: Dates:

___ Establish a shop drawing and samples review and approval procedure. (AIA Doc. 712) (See the Checklist at the end of the SHOP DRAWING section of this division of the manual.)

___ Hold a preconstruction meeting with all concerned parties, including the client, prime contractor(s), subcontractors, appropriate design firm staff members and management, and consultants.

___ Review the construction plan and overall contractual responsibilities of contractor(s), client, and design firm. Send any necessary clarification memos.

___ Send memos regarding interpretations, special instructions, modifications, and clarifications, as identified in the preconstruction meeting and any subsequent meetings or calls.

___ Confirm the construction start date, estimated days of construction, and the estimated date of substantial completion.

___ Confirm the client's approval of the construction schedule.

___ Confirm that all necessary insurance is in force prior to starting construction.

___ Confirm regulatory agency knowledge and approval of the scheduled construction start time.

___ Send the Notice to Proceed to the contractor.

___ Send copies of the Notice to Proceed to all concerned parties such as regulatory agencies, insurance representatives, testing companies, etc.

CONSTRUCTION FIELD OBSERVATION   (AIA Doc. G711)

Checkmark each item to be done and cross out the check when completed. Mark with a -- if an item is not to be done. If an item is in doubt, mark with question mark and add a note of what to do to resolve the question. By: Dates:

See THE GUIDELINES CONSTRUCTION ADMINISTRATION MANUAL, for construction inspection checklists.

___ Establish an assignment schedule and record log for periodic field observations and photo surveys of the construction site. Key construction phases requiring field observation include:

    ___ Preconstruction.
        ___ Jobsite construction layout.
        ___ Materials and storage.
        ___ Utilities and appurtenances.

    ___ Earthwork.
        ___ Excavation.
        ___ Grading.
        ___ Filling.

Add notes as needed after each checklisted task, such as:
Initials of who is to do it, when it's to start, when to review, who to coordinate with, and when it's to be finished.

Phase 7: Construction Contract Administration 6

CONSTRUCTION FIELD OBSERVATION (AIA Doc. G711) continued

Checkmark each item to be done and cross out the check when completed. Mark with a -- if an item is not to be done. If an item is in doubt, mark with question mark and add a note of what to do to resolve the question. By: Dates:

- ___ Foundations.
    - ___ Layout.
    - ___ Formwork.
    - ___ Pour.
    - ___ Basement waterproofing.

- ___ Concrete.
    - ___ Reinforcing.
    - ___ Formwork.
    - ___ Placement.
    - ___ Finishing.

- ___ Framing.
    - ___ Layout.
    - ___ Materials and storage.
    - ___ Erection.

- ___ Masonry.
    - ___ Layout.
    - ___ Materials and storage.
    - ___ Erection.
    - ___ Finishing.

- ___ Closing in.
    - ___ Roofing.
    - ___ Flashing.
    - ___ Temporary enclosures.
    - ___ Windows and doors.

- ___ Mechanical--Plumbing.
    - ___ Layout.
    - ___ Equipment and fixtures.
    - ___ Installation.
    - ___ Tests.

- ___ Mechanical--HVAC.
    - ___ Layout.
    - ___ Equipment and fixtures.
    - ___ Installation.
    - ___ Tests.

Add notes as needed after each checklisted task, such as:
Initials of who is to do it, when it's to start, when to review, who to coordinate with, and when it's to be finished.

Phase 7: Construction Contract Administration 7

CONSTRUCTION FIELD OBSERVATION (AIA Doc. G711) continued

Checkmark each item to be done and cross out the check when completed. Mark with a -- if an item is not to be done. If an item is in doubt, mark with question mark and add a note of what to do to resolve the question.   By: Dates:

___ Electrical.
    ___ Layout.
    ___ Equipment and fixtures.
    ___ Installation.

___ Finishes.
    ___ Walls and flooring.
    ___ Equipment installation.
    ___ Paint, hardware, fabrics, furnishings.
    ___ Landscaping.

___ Request photos from consultants and contractor, as appropriate to maintain a complete photo record of construction.

___ Establish construction photo standards of size; inclusion of chalkboard to show each photo date, time, and location; inclusion of scale rods where necessary to identify sizes of elements; and a site key map showing camera positions.

___ Establish a format for tape recorded construction site surveys and punchlisting.

___ Create a Field Observation Report format for observers from the prime design firm and consultant's representatives. Include space for:

    ___ Date, time, and duration of the site visit.
    ___ Names and titles of all parties dealt with during the visit.
    ___ Reason for the visit--scheduled routine visit, test observation, inspection request, etc.
    ___ Extent of the observations.
    ___ The condition of the work being observed. Particularly the condition compared to previous observations and previous Field Orders, Change Orders, and prior Field Observation Reports.

___ Create a calendar of special construction events and tests to be witnessed by design firm and/or consultant firm representatives.

___ Confirm that consulting firm and prime design firm representatives understand they must write a Project Field Observation Report after each site visit.

___ Review all Project Field Observation Reports, and submit a copy of each to the contractor's field representative. Include notices of any modifications required by observations in the reports.

___ Submit copies of Field Observation Reports, modifications, and construction schedule changes to the client.

___ Confirm that each Field Observation Report and corresponding construction photos (labeled and dated) are properly filed in the Construction Contract Administration Manual.

Add notes as needed after each checklisted task, such as:
Initials of who is to do it, when it's to start, when to review, who to coordinate with, and when it's to be finished.

Phase 7: Construction Contract Administration 8

FIELD ORDERS

Checkmark each item to be done and cross out the check when completed. Mark with a -- if an item is not to be done.
If an item is in doubt, mark with question mark and add a note of what to do to resolve the question. By: Dates:

___ Start a Field Order Log in the Project Construction Administration Manual. Note all supplemental drawings required for clarifications, interpretations, and revisions. (AIA Doc. G708)

___ Establish a procedure for noting field orders that affect the construction cost or time. Provide a cross reference system to assure that all such field orders are translated into change orders for approval by the client.

___ Establish a notation procedure for recording extra design and/or drawing services required due to changes initiated by the client.

___ Establish a file or log to verify that all negotiated change orders are approved and signed by the client.

CHANGE ORDERS

Checkmark each item to be done and cross out the check when completed. Mark with a -- if an item is not to be done.
If an item is in doubt, mark with question mark and add a note of what to do to resolve the question. By: Dates:

These and other items are included in QUOTATION REQUESTS AND CHANGE ORDERS.

___ Start a Change Order Log in the Project Construction Administration Manual. Note all supplemental drawings required for clarifications, interpretations, and revisions. (AIA Doc. G701, G701A, G709)

___ Review contractor Proposal Requests for changes with price quotations. (AIA Doc. G709)

___ Record extra design and/or drawing services required due to client-initiated changes.

TESTS

Checkmark each item to be done and cross out the check when completed. Mark with a -- if an item is not to be done.
If an item is in doubt, mark with question mark and add a note of what to do to resolve the question. By: Dates:

___ Start a Tests Log in the Project Construction Administration Manual. List all tests required for the job, and note their approximate likely dates according to the current construction schedule.

___ Confirm that required tests are being requested properly by the contractor.

___ Review all test results. Distribute test results to the appropriate design firm department heads and consultants so they can review test results that pertain to their work.

___ Confirm in writing that the contractor has responded properly when notified of deficiencies revealed in test results.

Add notes as needed after each checklisted task, such as:
Initials of who is to do it, when it's to start, when to review, who to coordinate with, and when it's to be finished.

Phase 7: Construction Contract Administration 9

CERTIFICATES FOR PAYMENT

Checkmark each item to be done and cross out the check when completed. Mark with a -- if an item is not to be done. If an item is in doubt, mark with question mark and add a note of what to do to resolve the question. By: Dates:

___ Start a Certificates for Payment record in the Project Construction Administration Manual. (AIA Doc. G702, G703, G722, G723)

___ Review applications for payment and compare each against:

    ___ Field Inspection Reports.
    ___ Construction photo records.
    ___ Previous project schedules.
    ___ Test reports.
    ___ Client observations.
    ___ Consultant observations.
    ___ Retained percentage.
    ___ Potential claims.
    ___ Substitutions.
    ___ Change orders.
    ___ Contractor's Schedule of Values.
    ___ Vouchers and receipts of payments to subcontractors and suppliers.
    ___ Materials paid for but not installed--storage and insurance.
___

___ Notify the contractor of any disapprovals or requests for further information to validate the application for payment.

___ Send the Certificate of Payment to the client, with a copy to the contractor. Confirm that the client has received it. (AIA Doc. G702)

AT 50% COMPLETION

Checkmark each item to be done and cross out the check when completed. Mark with a -- if an item is not to be done. If an item is in doubt, mark with question mark and add a note of what to do to resolve the question. By: Dates:

___ Update all project records.

___ Confirm that the contractor has complied with all field orders and change orders.

___ Review the contract sum retainage for possible reduction. If applicable, review the contractor's Consent of Surety.

___ Confirm that the monthly or periodic payments from the client to the design firm are up to date.

Add notes as needed after each checklisted task, such as:
Initials of who is to do it, when it's to start, when to review, who to coordinate with, and when it's to be finished.

Phase 7: Construction Contract Administration 10

CONSTRUCTION REPRESENTATION

Checkmark each item to be done and cross out the check when completed. Mark with a -- if an item is not to be done. If an item is in doubt, mark with question mark and add a note of what to do to resolve the question. By: Dates:

___ Establish a checklist and action calendar for full-time project representative(s). See AIA Document B352 for typical duties and limitations of project representatives.

CONSTRUCTION INSPECTION

Checkmark each item to be done and cross out the check when completed. Mark with a -- if an item is not to be done. If an item is in doubt, mark with question mark and add a note of what to do to resolve the question. By: Dates:

Note that items pertaining to inspection prior to substantial completion are noted here and also under SUBSTANTIAL COMPLETION AND PROJECT CLOSE-OUT.

___ Create a calendar of construction inspections. Determine inspections requiring the presence of design firm and/or consultant representatives.

___ Identify regulatory agency representatives who must inspect the building before occupancy. Schedule when they are to be notified of substantial completion.

___ Acquire regulatory agency inspection dates and notify the client and contractor.

___ Assemble copies of small-scale plans, approvals, test results, and any other data required to assist regulatory agencies in their inspections.

___ Review the inspection results of regulatory agencies, and obtain a punch list of required corrective actions.

SUPPLEMENTAL DOCUMENTS

Checkmark each item to be done and cross out the check when completed. Mark with a -- if an item is not to be done. If an item is in doubt, mark with question mark and add a note of what to do to resolve the question. By: Dates:

___ Prepare a Supplemental Drawing Log in which to record all new documents prepared for interpretation and clarification, and for recording project changes during construction. (AIA Doc. G710)

___ Prepare a log in which to confirm reproduction and delivery of all drawings and specifications that show design and schedule changes.

___ Schedule check times for confirming that supplemental drawings and other documents that show changes and clarifications have replaced obsolete documents.

Add notes as needed after each checklisted task, such as:
Initials of who is to do it, when it's to start, when to review, who to coordinate with, and when it's to be finished.

Phase 7: Construction Contract Administration 11

QUOTATION REQUESTS AND CHANGE ORDERS

Checkmark each item to be done and cross out the check when completed. Mark with a -- if an item is not to be done. If an item is in doubt, mark with question mark and add a note of what to do to resolve the question.   By: Dates:

___ Establish a Change Order Log, as noted on page 7A-14.

___ Establish a file or log in which to verify that all approved change orders are distributed to the contractor, accounting department, and all concerned consultants and departments.

___ Establish a file or log in which to verify that whenever there is an increase in the contract sum, you obtain copies of written consent provided by the surety company to the contractor.

___ Establish a routine schedule for filing copies of change orders and relevant drawings and photos in the Change Order section of the Construction Contract Administration Manual.

PROJECT SCHEDULE MONITORING

Checkmark each item to be done and cross out the check when completed. Mark with a -- if an item is not to be done. If an item is in doubt, mark with question mark and add a note of what to do to resolve the question.   By: Dates:

___ Schedule site visit dates specifically for determining the contractor's time schedule performance.

___ Schedule periodic memos for the client on the contractor's time scheduling performance. Note changes in schedule and any anticipated changes in the final occupancy date.

___ Make note in the time scheduling record to forewarn the client when evidence appears that liquidated damages may be required for slow performance.

CONSTRUCTION COST ACCOUNTING

Checkmark each item to be done and cross out the check when completed. Mark with a -- if an item is not to be done. If an item is in doubt, mark with question mark and add a note of what to do to resolve the question.   By: Dates:

___ Establish a file or log for evaluating Applications for Payment. (AIA Doc. G702, G702A)

___ With each Application and Certificate for Payment provide an accounting of:

    ___ Previous payments that have been approved.
    ___ Total of retainage withheld.
    ___ Cost effect of change orders.
    ___ Construction expenses compared with the original Schedule of Values.
    ___

___ Review and evaluate contractor expense records for cost-plus contracts.

___ Submit monthly statements to the client. (AIA Doc. G802)

Add notes as needed after each checklisted task, such as:
Initials of who is to do it, when it's to start, when to review, who to coordinate with, and when it's to be finished.

Phase 7: Construction Contract Administration 12

SUBSTANTIAL COMPLETION AND PROJECT CLOSE-OUT

Checkmark each item to be done and cross out the check when completed. Mark with a -- if an item is not to be done. If an item is in doubt, mark with question mark and add a note of what to do to resolve the question. By: Dates:

___ Review the Notification of Substantial Completion and the contractor's list of remaining work to be corrected or finished. (AIA Doc. G704)

___ Confirm that regulatory agency representatives who must inspect the building before occupancy have been informed of substantial completion.

___ Acquire regulatory agency inspection dates and notify the client and contractor.

___ Assemble copies of small-scale plans, approvals, test results, and any other data required to assist regulatory agencies in their inspections.

___ Confirm that the client has completed all final required applications for utility services and easements:
    ___ Electricity.
    ___ Telephone.
    ___ Gas.
    ___ Water.
    ___ Sewer district.
    ___ Utility heat or cooling medium.
    ___ Special communications cables.
    ___ Sidewalk easements/traffic rights-of-way.

___ Request from the client a list of items to be corrected.

___ Confirm that all required test reports have been received from the contractor.

___ Review test reports as pertain to substantial completion.

___ Review the inspection results of regulatory agencies and obtain a punch list of required corrective actions.

___ Do a field inspection of the project to confirm substantial completion. Refer to a Construction Administration Checklist of items to confirm components of work as correct or incorrect.

___ Prepare a punch list of remaining work to be repaired or completed. Include photographs as required for clarification.

___ Review a possible reduction of the contractor's retainage.

___ Inspect to confirm compliance with the punch list.

___ When the project is judged to be substantially complete, prepare a Certificate of Substantial Completion. (AIA Doc. G704)

___ Identify and assemble job documentation the client needs to send to financing agencies.

___ Obtain from the client written approval of the Certificate of Substantial Completion, and written acceptance from the contractor.

Add notes as needed after each checklisted task, such as:
Initials of who is to do it, when it's to start, when to review, who to coordinate with, and when it's to be finished.

Phase 7: Construction Contract Administration 13

SUBSTANTIAL COMPLETION AND PROJECT CLOSE-OUT   continued

Checkmark each item to be done and cross out the check when completed. Mark with a -- if an item is not to be done.
If an item is in doubt, mark with question mark and add a note of what to do to resolve the question.   By: Dates:

___ Have the contractor submit construction records and documents:

    ___ Certificates of Inspection and approvals.
    ___ Bonds.
    ___ Project record drawings ("as-builts").
    ___ Guarantees and warranties for installed equipment and materials.
    ___ Supplies and equipment provided for maintenance of installed equipment and materials.
    ___ Project maintenance manual as specified.
    ___ Equipment operating instructions.
    ___ Keying schedule and master keys.
    ___

___ Review the close-out records for completeness.

FINAL CLOSE-OUT

Checkmark each item to be done and cross out the check when completed. Mark with a -- if an item is not to be done.
If an item is in doubt, mark with question mark and add a note of what to do to resolve the question.   By: Dates:

___ Have appropriate consultants review the close-out records and documents for approval.

___ Have the client review the close-out records and documents for approval.

___ Notify the contractor of inadequacies in the close-out records.

SEMIFINAL AND FINAL INSPECTION

Checkmark each item to be done and cross out the check when completed. Mark with a -- if an item is not to be done.
If an item is in doubt, mark with question mark and add a note of what to do to resolve the question.   By: Dates:

___ Review the contractor's written notice that all work is completed.

___ Do a field inspection of the project to confirm completion and confirm that clean-up is satisfactory.

___ Prepare a final field inspection report listing items still remaining to be repaired and/or completed.

___ Conduct additional inspections as required due to incompleteness of the work. Deduct the cost of additional inspections from the contractor's final payment.

Add notes as needed after each checklisted task, such as:
Initials of who is to do it, when it's to start, when to review, who to coordinate with, and when it's to be finished.

Phase 7: Construction Contract Administration 14

FINAL CERTIFICATE OF PAYMENT

Checkmark each item to be done and cross out the check when completed. Mark with a -- if an item is not to be done. If an item is in doubt, mark with question mark and add a note of what to do to resolve the question. By: Dates:

___ Review the contractor's application for final payment, and acquire affidavits, receipts, vouchers, releases, and waivers of lien as necessary to verify the contractor's payments to subcontractors and suppliers. (AIA Doc. G706, G706A, G707)

___ Verify that the correct retainage is noted, to deduct from the payment to contractor.

___ Review release of liens. Investigate, as necessary, to make sure subcontractors haven't been coerced into signing releases.

___ Review the Consent of Surety.

___ Review the Owner-Contractor Agreement, and confirm that all conditions have been met. (AIA Doc. G702, G706, G706A, G707)

___ Issue a semifinal Certificate for Payment, if necessary, prior to issuing a final Certificate.

___ Send the Final Certificate for Payment to the client, and send a copy to the contractor. (AIA Doc. G702)

IN-OFFICE CLOSE-OUT

Checkmark each item to be done and cross out the check when completed. Mark with a -- if an item is not to be done. If an item is in doubt, mark with question mark and add a note of what to do to resolve the question. By: Dates:

___ Provide any required assistance to the client in obtaining final regulatory agency approvals, Certificate of Occupancy, etc.

___ Acquire photos, graphics, and project statistics for inclusion in the office press releases and brochure.

___ Confirm any final public relations and marketing services to be provided as part of basic services or as newly-agreed, added services.

___ Assist the client with publicity, open house activities, tenant rental brochures, etc. If not part of the original service contract, negotiate these as supplementary or added services.

___ Assist the client in installation of furnishings and signage. If not part of the original service contract, negotiate this as a supplementary or added service.

___ Assist tenants in interior design and signage. If not part of the original service contract, negotiate this as a supplementary or added service.

___ Move documents to interim storage during the warranty period:
    ___ Project manual and original working drawings.
    ___ Record drawings and prints.
    ___ Shop drawings.
    ___ Project Record including correspondence files.

Add notes as needed after each checklisted task, such as:
Initials of who is to do it, when it's to start, when to review, who to coordinate with, and when it's to be finished.

Phase 7: Construction Contract Administration 15

IN-OFFICE CLOSE-OUT   continued

Checkmark each item to be done and cross out the check when completed. Mark with a -- if an item is not to be done.
If an item is in doubt, mark with question mark and add a note of what to do to resolve the question.   By: Dates:

___ Assemble, summarize, distribute, and file the main elements of project history:

      ___ Construction cost records.
      ___ Design firm costs and fees.
      ___ Change order record.
      ___ Evaluation of the prime contractor and subcontractors.
      ___ Evaluation of the consultants.
      ___ Construction detail jobsite feedback forms and final evaluation of key construction details.
      ___ Evaluation of the working drawings.
      ___ Evaluation of the specifications.
      ___ List of trouble spots for inclusion as cautionary notes in the office manual.

___

___ Store in accounting files:

      ___ Close-out records.
      ___ All project accounting records.

___

___ Store in computer files:

      ___ All small- and large-scope computer graphics.
      ___ Project planning, checklists, and calculations.
      ___ All job-related word processing text.
      ___ Evaluate computer-generated graphics for potential reusability. Code, catalog, and file accordingly.

___

___ Confirm that copies of all documents required for long term liability considerations are catalogued and stored and that originals are stored separately in a vault facility.

PAYMENT TO THE PRIME DESIGN FIRM

Checkmark each item to be done and cross out the check when completed. Mark with a -- if an item is not to be done.
If an item is in doubt, mark with question mark and add a note of what to do to resolve the question.   By: Dates:

___ Prepare the final statement for design services and the final statement for reimbursable expenses; submit statements and documentation to the client. (AIA Doc. F502)

Add notes as needed after each checklisted task, such as:
Initials of who is to do it, when it's to start, when to review, who to coordinate with, and when it's to be finished.

Phase 7: Construction Contract Administration 16
Shop Drawing Checking and Coordination

PROJECT MANAGEMENT CHECKLIST

Project Name/No: Notes by:

Dates Checked:

ADMINISTRATION

Checkmark each item to be done and cross out the check when completed. Mark with a -- if an item is not to be done. If an item is in doubt, mark with question mark and add a note of what to do to resolve the question. By: Dates:

___ Start and maintain a Shop Drawing and Sample Record to record all shop drawings received from the contractor. (AIA Doc. G712)

___ The shop drawing record or log should contain space for the following data:

  ___ Project identification.
  ___ Contractor.
  ___ Design firm ID number.

  For each submittal:

  ___ Data received.
  ___ Related specification section number.
  ___ Drawing or sample number.
  ___ Drawing or sample name.
  ___ Manufacturer or supplier.
  ___ Contractor/Subcontractor/Trade.
  ___ Number of items received and number referred or returned.
  ___ Date of referral or return.
  ___ Routing of referrals or returns.
  ___ Routing of copies to the client, field representatives, consultants, contractor, and files.
  ___ Action taken and date.
  ___ Scheduled date for resubmittal, if required.
  ___ Name of checker.
  ___

___ Review the General and Supplementary Conditions describing the relative responsibilities of the contractor and design firm in regard to shop drawings.

___ Review or make agreements with consultants regarding their responsibilities in checking shop drawings.

___ Create a list of anticipated shop drawings and a preliminary calendar schedule of when they should be received by the contractor and when submitted to the design firm.

___ Schedule periodic reviews of the shop drawing calendar for the purpose of reminding all parties involved of shop drawing needs.

Add notes as needed after each checklisted task, such as:
Initials of who is to do it, when it's to start, when to review, who to coordinate with, and when it's to be finished.

Phase 7: Construction Contract Administration 17
Shop Drawing Checking and Coordination

ADMINISTRATION   continued

Checkmark each item to be done and cross out the check when completed. Mark with a -- if an item is not to be done. If an item is in doubt, mark with question mark and add a note of what to do to resolve the question.   By: Dates:

___ Submit lists to the contractor of construction items that will require:

    ___ Shop drawings.
    ___ Product data.
    ___ Product and material samples.
    ___ Color samples.
    ___

___ Establish a transmittal form for conveying reasons for disapproval of a sample or show drawing. (Consider using a checklist such as the CHECKING AND APPROVING SHOP DRAWINGS AND SAMPLES checklist at the end of this section.)

___ Establish a tickler file or time log to track the location and progress of shop drawings and samples that are referred to others, items returned for correction and resubmittal, etc.

___ Notify or remind the contractor to maintain copies of all approved shop drawings, samples, and colors to provide to the client at the closeout of the project.

___ Schedule a check to reconfirm that copies of all approved items are being kept by the contractor for the client.

PROCESSING AND ROUTING SHOP DRAWINGS

Checkmark each item to be done and cross out the check when completed. Mark with a -- if an item is not to be done. If an item is in doubt, mark with question mark and add a note of what to do to resolve the question.   By: Dates:

___ When shop drawings arrive at the design firm for checking:

    ___ Fill in the "received" line of the Shop Drawing and Sample Record form.
    ___ Add the date received to the shop drawing.
    ___ Add the design firm's file number.
    ___ Add identifying and sequence number.
    ___ Check the shop drawing stamp for completeness. Be sure additional text on the stamp that explains the relative responsibilities of the design firm and the contractor conforms to the General and Supplementary Conditions of the contract.
        ___ If the drawing is to be checked by another party, add a routing stamp with destination and note of the date of transmission.
    ___

Add notes as needed after each checklisted task, such as:
Initials of who is to do it, when it's to start, when to review, who to coordinate with, and when it's to be finished.

Phase 7: Construction Contract Administration 18
Shop Drawing Checking and Coordination

CHECKING AND APPROVING SHOP DRAWINGS AND SAMPLES

Checkmark each item to be done and cross out the check when completed. Mark with a -- if an item is not to be done. If an item is in doubt, mark with question mark and add a note of what to do to resolve the question. By: Dates:

___ Establish a checking sequence and criteria such as:

    ___ Conformance of submitted item to specified item, model, and type.
    ___ Explanation of substituted items and conformance of substitutions to specified standards.
    ___ Conformance to required testing criteria such as ASTM, ANSI, ASME, or UL.
    ___ Conformance to specified performance or capacity i.e. quantified strength, voltage, pressure rating, etc.
    ___ Item submitted matches connecting materials, equipment, and fixtures.
    ___ Item submitted matches built-in related work and services such as electrical, plumbing hookup, etc.
    ___ Conformance to code.
    ___ Certificates of testing submitted as specified.
    ___ Conformance to required shape, form, and overall appearance.
    ___ Conformance to required size dimensions and tolerances.
    ___ Conformance of optional accessories to specifications.
    ___ Materials as specified:

        ___ Thickness or gauge.
        ___ Weight or density.
        ___ Standard or rating.
        ___ Finish.
        ___ Color.
        ___ Additional treatment.
        ___ Exposure to weather, moisture, light, heat, abrasion, chemicals, etc.

    ___ Fastenings:

        ___ Material.
        ___ Type.
        ___ Size.
        ___ Spacing.
        ___ Backing or connecting material.

    ___ Relationship to other construction:

        ___ Fit.
        ___ Maintenance access.
        ___ Noise consideration.
        ___ Fire consideration.
        ___ Safety consideration.
        ___ Cathodic or chemical reaction with other materials.
        ___ Effects of shrinkage, or thermal expansion and contraction.

Add notes as needed after each checklisted task, such as:
Initials of who is to do it, when it's to start, when to review, who to coordinate with, and when it's to be finished.

Phase 8: Postconstruction Administration 1

PROJECT MANAGEMENT CHECKLIST

Project Name/No: Notes by:

Dates Checked:

ADMINISTRATION - UPDATES AFTER CONSTRUCTION
Checkmark each item to be done and cross out the check when completed. Mark with a -- if an item is not to be done.
If an item is in doubt, mark with question mark and add a note of what to do to resolve the question.   By: Dates:

___ Confirm postconstruction services already agreed to and negotiate any additional services desired by the client. (AIA Doc. B162)

___ Back check and clear left over tasks from the PHASE 7: CONSTRUCTION CONTRACT ADMINISTRATION checklist.

___ Bring project records up to date by recording all pertinent discussions and decisions from the previous phase that haven't yet been recorded.

___ Update contact names, phone numbers, addresses, remarks, etc. in the Project Directory.

___ Input Project Directory updates into the office-wide Project Directory data base.

DISCIPLINES COORDINATION AND DOCUMENT CHECKING
Checkmark each item to be done and cross out the check when completed. Mark with a -- if an item is not to be done.
If an item is in doubt, mark with question mark and add a note of what to do to resolve the question.   By: Dates:

___ Check that consultants update their portion of record drawings and related documents.

___ Coordinate consultant's observations of the contractor's remedial work on:
    ___ Sitework and site drainage.
    ___ Landscaping.
    ___ HBAC.
    ___ Plumbing.
    ___ Lighting.
    ___ Communications.
    ___

AGENCY CONSULTING, REVIEW, AND APPROVALS
Checkmark each item to be done and cross out the check when completed. Mark with a -- if an item is not to be done.
If an item is in doubt, mark with question mark and add a note of what to do to resolve the question.   By: Dates:

___ Assist the client in obtaining any delayed final permits or certificates.

Add notes as needed after each checklisted task, such as:
Initials of who is to do it, when it's to start, when to review, who to coordinate with, and when it's to be finished.

Phase 8: Postconstruction Administration 2

OWNER-SUPPLIED DATA COORDINATION

Checkmark each item to be done and cross out the check when completed. Mark with a -- if an item is not to be done. If an item is in doubt, mark with question mark and add a note of what to do to resolve the question. By: Dates:

Also see POSTCONSTRUCTION - ADDITIONAL SERVICES.

___ Assist the client in selection and installation of client supplied furnishings, fixtures, and equipment.

MAINTENANCE AND OPERATIONAL PROGRAMMING

Checkmark each item to be done and cross out the check when completed. Mark with a -- if an item is not to be done. If an item is in doubt, mark with question mark and add a note of what to do to resolve the question. By: Dates:

___ Coordinate maintenance and instruction meetings between the client and the equipment manufacturers' representatives.

___ Assist the client in assembling and building operations and maintenance manual(s).

START-UP ASSISTANCE

Checkmark each item to be done and cross out the check when completed. Mark with a -- if an item is not to be done. If an item is in doubt, mark with question mark and add a note of what to do to resolve the question. By: Dates:

___ Assist the client in training operational and maintenance staff members in correct work scheduling and procedures.

___ Schedule and coordinate corrective work by the contractors.

RECORD DRAWINGS

Checkmark each item to be done and cross out the check when completed. Mark with a -- if an item is not to be done. If an item is in doubt, mark with question mark and add a note of what to do to resolve the question. By: Dates:

___ Obtain certification from contractors of changes made by them during construction.

___ Identify all concealed systems installed by contractors.

___ Prepare or review record drawings to confirm that all notable changes made during construction have been accurately recorded.

___ Send record drawings and related data to the client and to any others as directed by the client.

Add notes as needed after each checklisted task, such as:
Initials of who is to do it, when it's to start, when to review, who to coordinate with, and when it's to be finished.

Phase 8: Postconstruction Administration 3

WARRANTY REVIEW

Checkmark each item to be done and cross out the check when completed. Mark with a -- if an item is not to be done. If an item is in doubt, mark with question mark and add a note of what to do to resolve the question. By: Dates:

___ Schedule inspection of materials and equipment prior to the expiration of warranties.

___ List all corrective work required of the contractor.

___ List all corrections, repair, or replacements required of equipment and materials manufacturers.

___ Assist the client in overseeing corrective and replacement work.

POSTCONSTRUCTION EVALUATION AND POST-OCCUPANCY SURVEY

Checkmark each item to be done and cross out the check when completed. Mark with a -- if an item is not to be done. If an item is in doubt, mark with question mark and add a note of what to do to resolve the question. By: Dates:

___ Interview maintenance and operational personnel for evaluation of the building design, systems, materials, and equipment.

___ Acquire operation and maintenance costs from the client.

___ Conduct field observation by design, production, and A/E managerial staff to evaluate the functioning of design, planning, systems, materials, and equipment.

___ Photograph visible construction problems. Make copies for jobsite feedback to the standard construction detail system.

POSTCONSTRUCTION - UPDATING THE OFFICE DATE BASE

Checkmark each item to be done and cross out the check when completed. Mark with a -- if an item is not to be done. If an item is in doubt, mark with question mark and add a note of what to do to resolve the question. By: Dates:

___ Start or continue an office DESIGN AND PLANNING DATA BASE. Create a list of design features of the project with notes on how those features may be improved and applied in future projects. (A data base may be kept as a file folder system, in binders, in computer storage, or in any combination of these.)

___ Identify plan features such as repetitive rooms or suites, utility spaces and special function rooms, etc. that might be adapted from this project for use in future projects. Include these in the DESIGN AND PLANNING DATA BASE.

___ Start or continue an OFFICE MANAGEMENT DATA BASE. Create a list of management elements, procedures, forms, checklists, etc. that worked well on this project. List those that didn't work as expected with notes as to how they can be improved.

___ Start or continue a TECHNICAL DATA BASE. Create a list of technical features - details, specification sections, construction procedures, etc. that were especially successful in this project.

Add notes as needed after each checklisted task, such as:
Initials of who is to do it, when it's to start, when to review, who to coordinate with, and when it's to be finished.

Phase 8: Postconstruction Administration 4

POSTCONSTRUCTION - UPDATING THE OFFICE DATE BASE

Checkmark each item to be done and cross out the check when completed. Mark with a -- if an item is not to be done.
If an item is in doubt, mark with question mark and add a note of what to do to resolve the question. By: Dates:

__ Review all construction details created for this project. Identify and file those that should be made a part of the office standard detail system. Identify and file those that should be filed as design reference details.

__ Review this checklist manual for special content you have added that you believe should be a standard part of future checklists.

__ Make a master copy of the PROJECT MANAGEMENT TASK MODULES MANUAL with your updates and revisions for use on future projects.

POST-CONSTRUCTION - MARKETING

Checkmark each item to be done and cross out the check when completed. Mark with a -- if an item is not to be done.
If an item is in doubt, mark with question mark and add a note of what to do to resolve the question. By: Dates:

__ Obtain interior and exterior photographs for publicity submittals and the office brochures.

__ Obtain client testimonials regarding the working relationship with the design firm.

__ Compile finished building statistics: Square footage, final construction costs, construction schedule, speed of tenant rentals, owner return on investment, etc. for publicity and brochure text.

__ Update the office resumes with data on consultants and staff who participated in the project.

__ Prepare a summary of the primary programming and predesign problems and needs and how the final design addressed those needs.

__ Locate and catalog all documents that are reusable in future publication and marketing:
  __ Models and presentation drawings.
  __ Design sketches.
  __ Slides, photos, and video tapes.
  __ Base sheets and overlays.
  __

POST-CONSTRUCTION - FACILITIES MANAGEMENT

Checkmark each item to be done and cross out the check when completed. Mark with a -- if an item is not to be done.
If an item is in doubt, mark with question mark and add a note of what to do to resolve the question. By: Dates:

__ Negotiate tenant space planning, interior design, and remodeling contracts.

__ Negotiate facilities management services.

__ Negotiate maintenance scheduling services.

__ Negotiate building tenant rental management services.

Add notes as needed after each checklisted task, such as:
Initials of who is to do it, when it's to start, when to review, who to coordinate with, and when it's to be finished.

Phase 9: Long-Range Marketing Planning 1

LONG RANGE MARKETING PLANNING

Checkmark each item to be done and cross out the check when completed.
If an item is in doubt, mark with question mark and add a note of what to do to resolve the question. By: Dates:

This list of management tasks is for group consideration or use it for introspection if you're designing your own personal long term marketing program. This exercise has been widely used to help get track with an effective marketing program.

___ Identify the primary life desires, needs, and professional strengths of the office principal(s). List three to six items in each category on a separate page.

___ Based on life needs, identify the personal long-term professional goals of the office principal(s). List three to six primary and secondary goals per principal. (Use a separate sheet if necessary.)

___ Identify the office's long-term goals/objectives best suited to serve the aggregate of the principal(s) personal goals. List three to six specific goals/objectives. Find one that best summarizes overall office intentions and purpose.

___ Identify the client markets best suited to the office's long range goals. List up to six markets and identify the best.

___ Identify and list optimal office characteristics to meet the goals. Create separate lists as necessary.

    ___ Size.
    ___ Specialization.
    ___ Income and income growth.
    ___ Office financial structure.
    ___ Managerial structure and style.
    ___ Location.
    ___ Affiliations.

___ Identify and list office weaknesses and trouble spots. List specific steps with a timetable for remedying the defects. Create separate lists as necessary.

    ___ Coordination.
    ___ Long range planning.
    ___ Financial management.
    ___ Fee collection.
    ___ Meeting deadlines and schedules.
    ___ Obsolete methodologies.

___ Identify your office's primary structural and physical needs. Create separate lists as necessary.

    ___ Financial.
    ___ Staffing.
    ___ Communication.
    ___ Organization.
    ___ Environment.

Add notes as needed after each checklisted task, such as:
Initials of who is to do it, when it's to start, when to review, who to coordinate with, and when it's to be finished.

Phase 9: Long-Range Marketing Planning 2

LONG-RANGE MARKETING PLANNING continued

Checkmark each item to be done and cross out the check when completed.
If an item is in doubt, mark with question mark and add a note of what to do to resolve the question. By: Dates:

___ Identify an appropriate focal point of design service. Create separate lists as necessary.

    ___ Aesthetic.
    ___ Administrative/managerial.
    ___ Technical.

___ Identify a client market appropriate to the office's strengths and needs. Create separate lists.

    ___ A profile of acceptable client types and characteristics.
    ___ Region(s).
    ___ Names/prospect directory.

___ Establish targets of how many new clients to contact and to sign within three, six, twelve, and eighteen months.

___ Establish a client prospect contact calendar and schedule.

    ___ Previous client contact program.
    ___ Cold call contacts.
    ___ Letters of inquiry.
    ___ Trade association contact.
    ___ Response system for RFP's and referral/lead sources.
    ___ Prospect visits and brochure drop-off.
    ___ Prospect diagnostic review or project audit.
    ___ Call backs.

___ Alternative services to review with clients that lead to projects.

    ___ Property utilization audit.
    ___ Project permit acquisition planning.
    ___ Project feasibility study.
    ___ Project audits of previous projects rejected for financing and/or regulatory agency permit.

___ Establish a publicity program to increase your firm's recognition among prospective clients. Identify who can contribute in each area of publicity.

    ___ Brochure design.
    ___ Talks, lectures, exhibits and other educational events.
    ___ Articles and other publications.
    ___ Informational newsletter.
    ___ Services of marketing consultant for all marketing.
    ___ Services of a publicist for publication in newspapers and magazines.
    ___ Submittals for awards.

___ Schedule periodic progress reviews and updates of the marketing program.

Add notes as needed after each checklisted task, such as:
Initials of who is to do it, when it's to start, when to review, who to coordinate with, and when it's to be finished.